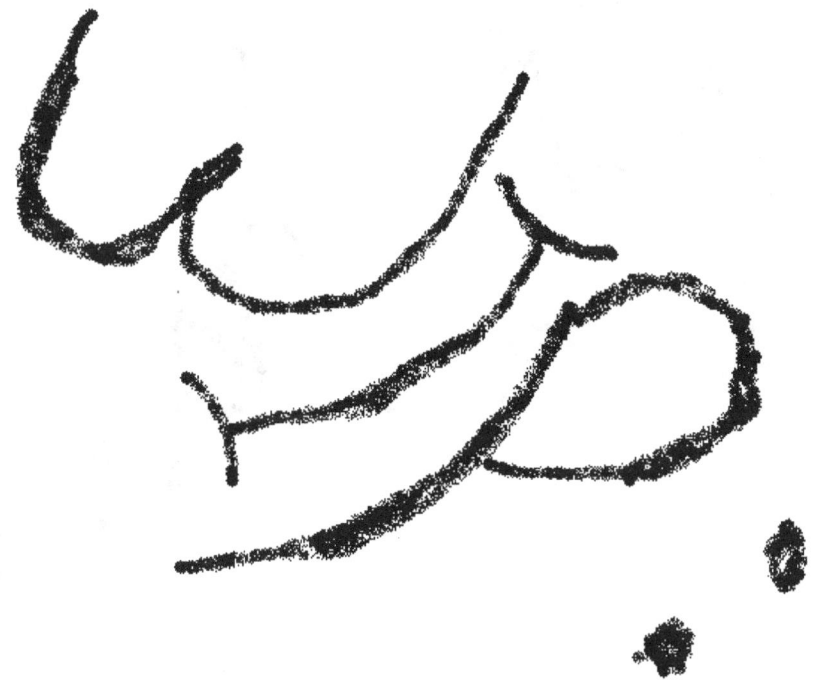

and have

deal for

you?

for

Ya know, it really did seem like this went on forever. Keep runnin' into problems.

But now I can add a coupla more things:

The actual e-mail address is on the back cover. I F#<%łd up on the one that's later in the book.

You're goin' havta buy T... H... E... to understand the next two:

Stephanie has a genius level IQ too. 142. But she ain't got no education, so she sells tobacco. She seems happy doin' it.

The crazy dude usin' his emergancy brake to slow down for curves: First name: Jamey.

And the guy who owns the tobbacco shop would rather lose money than overcharge his customers. Ain't many around like that any more.

Maybe I really will write a 3rd book tellin' ya all about how much fun I had doin' this. I think the cops may have been after me after a bit of road=rage.

Time is a perception. Don't ask me, ask Einstein.

Goin' keep my fingers crossed one more time.

Oh, yeah. One more thing:

Some one from Indiana who should be famous but ain't:

Damn, thought his name would be on the album but it ain't. First name Dan. Maybe Don.

Played in a band from Muncie that first went by the name of The Choosen Few, then: Faith, finally: the Faith Band.

Played the saxophone with one hand and the keyboard with the other. At the same time!

He's dead now. Damn! So's Kurt. Double Damn!! I think I got it this time. So:

No, I'm still not certain I got the fonts embedded. But Fuk It.

That's all for now, Folks.

(A^C ((BUT if I can remember how I did this, Ican use it for E = MC squared!)

Damn! I'm tryin' to use a PDF Editor and I don't know how to use it!)

(Actually it's rougher than I i originally thought)
(I've probably misspelled Vonnegut.)
  (Sorry about that Kurt. I admire you a lot.)

goin'to lunch now.
    with DAd.

(Actually, I didn't. I just googled him. Whew.)

(Wasn't sure about signifyin' titles.)

Uriah Heep: "Demons
    and
      Wizards"
       We'll never be alone.

(But I'm gettin' impatient.)

(And they've got proofreaders at the publisher's.)

(Honest-to-God, folks: I'm laughin' out loud, laughin' out loud

This is goin' to be a little rough.

(And here's somrthin' else that happened.
They said they took wpd files.
And they couldn't open mine!
This version must be really old!)

Don't know WordPerfect that well.

(Thought
shouldn't put in
real name.)

(He put this in there when he e-mailed this thingy to Mary at ------ Publishers.

Wanted to warn her that he was no professional writer.

It was about 100 pages at the time.

Who knows how long it will be when he finishes this **damn** thing?)

(April Wine playin'

"21st Century Schizoid Man".

It's a cover.)

(Maybe, I shouldof charged more
Than 10 dollars for this
DaMN thingy.)

(Pound them knees!)

at what I've just done.)

(I've been wlkin' in circles, thinkin': WHAT THE HELL I'M I DOIN"!)

damn, see Karen, I am a lousy typist, It's just that usually I fix my mistakes.

(LOL, this editor is kinda neat. I can stick text where ever I want it!)

(It's only a 14 day free trial, so I gotta hurry.)

(I just put on: " Break on Through to The Other side")

(The Doors.)

(Jim Morrison singin')

(I'm tryin' Jimmy.)

1

WELCOME TO MY WORLD:

Come on in.

The waters are fine.

By

Mike Barden

Dedicated to:

Lucille Mae Wilson Barden

And

William J. Barden

Foreword from the author:

To all the people I've ever met.

Hello.

How ya doin'?

I'm doin' Okay.

Thanks

for helpin' me

Become

who I am.

I couldn't have done it without ya!

don't cry

Let me do it fo ya.

("It's not sane". Blind Melon this time. And this page wasn't empty when I scrolled through the

first time.)

The American Dream.

Work.

Struggle.

Pay them bills!

How Strange!?

{Don't give up yet. You'll hear from Mary.}

{OK}

(Actually, they put me in touch with Courtney.)

Forgive him for that.

He's actually **proud** to be American.

But sometimes he wonders about that dream.

6

time to pay those bills Actually, it took about a week before I actually got around
to payin' those bills. I kept wrtin' instead.

(Originally, time to pay those bills was at the bottom of the preceding page.)

(But it got moved when he translated this thingy out to Word.)

(So he could show what he was doin' to someone else.)

(And he kinda likes it on it's own page.)

(Time to eat lunch with Dad before goin' to work.)

(Got worried that this was another damn empty page.)

(And maybe I'll mail those bills.)
(A couple of them might be late.)

You Know,

I've had my mind elsewhere,

And

It's had me elsewhere,

For a while now.

And

I

Haven't

Paid much attention

To current affairs.

I'm

Watching

<u>60 Minutes</u>

And

I

Go:

I **really** like this guy!

I voted for Obama but that was just before the madness began.

Maybe **he** will be able to turn this whole thing around!

A suggestion from the author before undertaking this endeavor:

He really kinda let his subconscious be his guide on this. Let it take him where ever it wanted to go. It does jump around a lot, especially at the beginning. He mellowed out some by the end. But it jumps around some even then.

So, the best way to approach this might be:

Think of this as a poem. You don't necessarily try to make sense of every word when you first read a poem. You just kinda read it and let it all soak in.

Yeah, think of this as some kind of poem.

The
Surrender yourself to it! Crystal Method: "Trip like I do"

"Turn your head"

(Damn! Translatin' it to the PDF editor messed with justification again

It was the end of his days. He felt it coming. Not psychic, he just knew.

His leg was numb. His chest crushing, heart pounding, light-headed, dizzy. Something was coming. That he knew.

Much knowledge would be lost from a lifetime of learning. But it could not be helped.. Others had learned from his ways.

Cramps in his hands.

The Universe was a series of infinite spherical loops. Expanding, contracting, like a heartbeat, music.

He had dreams. Some good, some bad, all real. Walking down a nice suburban street. White picket fence. He enters and it's a carnival fun house! Rooms distorted. He wakes, crying. The bed is like concrete. His hands entwined in the hair of his mother. He is a scared child, who'll never grow up. The dark will be his friend.

His favorite movies? It depended on his mood. 2001: A Space Odyssey. The concept infinite; the dialogue mundane, everyday. He didn't cry that much anymore but A Wonderful Life made him cry. He went to see A Clockwork Orange, twice, with friends. They came out disturbed by the violence. He came out Singing in the Rain. The juxtaposition, the irony of that scene. That is what he loved most, irony. But he was losing his sense of it. The irony.

He had also lost his love of learning. Dead-end jobs will do that to you. You want to feel secure. Your place in the world. But that wasn't his. What was? He was still trying to find that place.

Security. He wasn't ready to die quite yet.

Intestines churning.

His mind races, then crawls.

In life, he was not that good at keeping up with old friends. His life kept changing, mutating. Not sure what they had in common anymore. Maybe in death he could once again say: Hello. How ya doing. Thanks for being there when I needed you.

Muscles twitch. Not time yet.

Read all of Anthony Burgess's books. Strange M/F. Now that he thinks about it, M/F was a title of one of his books.

Singing in the Rain.

He first found himself at Arcosanti, the city of the future, city built in the image of man, out in the Arizona desert. He was a god. Arrogant? No, he thought that everyone was their own gods, shaping their own Universes. Saw Mary Poppins, again, while out there. A film for the revolution brought to you by Disney. Up the Establishment and fly. He had been on a bridge, deep in thought. The idea of flying through the air, breath-taking. He forgot the logical conclusion: the ground. He found himself in the middle of the bridge, not the edge. Had instinct known that if he had been at the edge...

He had considered suicide many a time, in the sort of maybe they'll miss me when I'm gone kinda way. It seemed logical. Never had the emotional drive, the impulse, till he went through, what he later found out to be, discontinuation syndrom. Scary. No one told him anti-depressants can be addictive. No one tells a lot.

Infinite spherical loops.

But I'm not goin' to try to fix it this time.)

Endless paperwork. A receipt for everything. A recipe for everything? Everything is relative. Once heard the genes we all shared referred to as junk. Seemed harsh. After all, they are what we all have in common. Maybe some secret is hidden in there, mutating.

More pain.

Made an unsuccessful attempt at getting a graduate degree in Computer Science. Wrote a paper for a Computer Hardware class he titled: <u>Random Order and Infinite Variability</u>. Had little to do with computer hardware. Still got a B. He didn't know whether that said more about the teacher or himself. The teacher even made a suggestion for further proof, some mathematical equation he had to look up. He couldn't see how it applied.

Someone said it before: Love having written, Hate writing. The memories can hurt.

Intestine collapsing.

About time to go back to that dead-end job. He had been off for a while, dealing with the pain, the stress. But the bills pile up.

More pain.

Loops, infinite, spherical.

A infinite number of monkeys at a infinite number of typewriters, writing Shakespeare.

Time to go back to work. Not ready to die quite yet. He'll be back later.

<u>Slaughter House 5</u>

Kurt Vonnegut. Another strange M/F. Aliens watching. Observing. The Observer alters the Observed which alters the Observation which alters the Observer which...

Hands entwined in his mother's hair.

Infinite,

Spherical,

Loops.

This could go on for ever.

Another suggestion from the author before goin' any further:

He tried to show this thingy to a co-worker.

It seemed that she might have been through the madness.

So, he thought:

She might get it!

He hadn't written that much, yet.

He was havin' problems gettin' what he did have printed out. There's a bunch of reasons for that, so he won't try to explain. (Has to do with file formats.)

So, all he had then were the first couple of pages.

Her response:

I had to read the thing a couple of times.

You can really see your mind jump around.

(She said it like it was a bad thing,)

It is well written, though.

So:

He really did try to let his subconscious be his guide on this.

Let it led him where ever it wanted to go.

He's starting to think ahead a little bit more now.

(Dad just called. I keep getting interrupted!)

(The Who: I know that you have because there's magic in my eyes.)

{I can see for miles and miles.}

Maybe, now I'm gettin' too clever!

15

He does kinda throw some things out there in the beginning.

Poofin' the donuts?

He hada sorta Jonathan Winters, Robin Williams thing goin' on at the time.

It was his best way of explaining the process of putting the filling into the donuts.

The people he was with then found it kinda amusing.

They don't now.

Maybe that's because now he tries to use real words instead of makin' them up.

So, he stopped doin' that kinda thing.

The monkeys writing Shakespeare?

That he explains on page, let's see, last time he looked, it was on page 22. Who knows where it will be by the time he ends this thing?

(The Who said it first: You know my curse has ended.)

You know, at the time I'm wrtin' this, I don't know if this explanation will be in this book. It's in another file and and in another format and I'm not sure if I can merge them. And maybe it shouldn't be. It apparantly isn't a good enough explanation. I showed it to Karen and she said: I read it three times and I still don't understand what yor're talkin' about.

The Observer... thing.

Sorry about that, Karen. I was gettin' kinda proud of myself. I thought I had come up with a pretty good explanation and I fail;ed. I'm still workin' on it.

That he pretty much waits to the end to explain.

Anyhow.

I said somethin' about my mind racin' then crawlin'. You know what's happenin' there: I'm tryin' to find the middle ground. It races because I'm workin' on a problem. tryin' to connect the dots. when it crawls, I've kinda given up. I don't want to think anymore. But at the moment in the middle, there's this clarity.

Maybe, I am a machine after all.
I constantly need adjustin'!

The Universe...?

I'm debatin' about deletin' the above, because I'm afraid I've might of reveled(I'm goin' to leave spellin' mistakes alone to prove how bad I am at such things.) too much. And if I do,

To figure that one out, you're probably goin' to havta read the whole damn book.

I'm goin' leve this here and really screw with your minds. LOL

That's his best way of explaining that.

"A little evil inside": Disturbed again.

This I am definetly goin' to leave in:
Thank you Karen. You've been a real inspiration.

So,

"Prayer": Disturbed

16

pound them knees!

Again!

Think of this as some kinda poem.

Stop thinkin' about it and just keep reading.
Please! Got a little frustrated. Sorry about that.

this could go on for ever

Surrender yourself to it.

(I'm the gypsy, the acid queen. I'm guaranteed to break your heart!)
(He never did any acid. It scared him.)

{I'm free, and I'm waitin' for you to follow me.}
(Freedom tastes of reality)
{How can we follow?}

Just keep readin'

(The Who said it first: There ain't no cure for the summertime blues.)
{Actually, He thinks that's a cover song.)
(It was the Blues Magoos who said it first.)
{He thinks it was a cover song even then!}
(Johnny Something.)

Just keep readin' Please!

I feel like I have to keep going back and explain things I've all ready written

And

I need to finish this damn thing.

17

{This really, really could go on for ever. There are many paths to take.}

(He just put on Pink Floyd's <u>The Dark Side Of The Moon</u>. Take a time-out and put it on yourself and listen to the first couple of sentences right at the beginning. They're kinda hard to make out, you may not recognize them as actual words at first, but maybe then you'll understand WHY he's doin' this thin' here.)

(Pound them knees!)

(While you're at it: Listen to the whole damn CD. He is.)

II

loop

The pain had started out as a minor discomfort, deep in the upper abdomen, right at the boundary of abdomen and chest. It felt like muscle contracting abnormally. And it occurred when and only when he sat. He waited for a month, hoping that it would just go away. It didn't. Off to the doctor, figuring this would be easy. I know where it occurs, how it feels, when it occurs. Apparently, nothing quite fit that description.

There's that constriction in the chest again.

Gasps.

Light the pipe.

Discomfort in the anal region.

Pick up the paper.

Put it down.

A constant struggle. Id/Ego. He was once told by a woman that he had no ego. She said it like it was a bad thing. Eastern Philosophies consider loss of ego to be good. Western Philosophies consider ego necessary for success. A constant struggle.

Hands entwined in his mother's hair.

She died during this struggle with the pain. He was not there at the time, but had slept on the couch near her bed for a week or two. She seemed peaceful. Alzheimer's. A doctor once asked him how she was doing. Deep in his own pain and worries, he replied, angrily: she's off in her own world. How callous from one not normally callous. Sure, she was off in her own world a lot, but so was he! And there were times when he thought: she really hit the nail on the head. Cut through all the bullshit. Got to what was really important.

More pain.

Hope that she forgives him for that remark.

Hands entwined in her hair.
Infinite
Spherical
Loops.
This could go on for ever.

III

loop

He returns home from work for the first time in three months. Not as bad as he thought it would be. Forgotten that even dead-end jobs have their value. It was not his first dead-end job, just the first he thought he might retire from. No longer. It killed his spirit, the love of learning, his curiosity. Black Holes, particles coalescing to a single point, a single particle devoid of motion, devoid of time. That's what he imagined. That's what made him happy. But for now, that dead-end job would have to do.

More pain.

Hands entwined in his mother's hair.

Infinite

Spherical

loops.

His supervisor mentioned seeing Ravi Shankar on a TV program. He remembered playing with a sitar player who had played with Ravi Shankar. Jamming would be more accurate. It happened at a new-age, clothing-optional, hot spring resort in Northern California where he had gone when his destiny as a rich-and-famous movie director would not be fulfilled.

The journey is as important as the destination.

He use to pound along to the music everywhere. Tables at top-less bars. His knees.

Hands entwined in her hair.

Hands entwined in his mother's hair.

Hear her heart beat, the rhythm, the music.

A woman had lent him her drum, maybe African. He didn't know. Didn't care. It was something to pound on. High on psilocybin. (Before he goes farther: today's drug culture is very violent, the drugs destructive.) Without the psilocybin, he would not have had the guts to do what he did. Felt self-conscious being the center-of-attention. But if he could get into the zone, well, that's another thing entirely. Should be possible without the drugs. Become a child again. Don't lose that sense of awe.

Hands entwined in his mother's hair.

Infinite

spherical

loops.

This could go on for ever.

The house band had taken a break. The lead guitarist-and-singer had played for Little Richard. He took the drum, sat on the stage, and pounded away, oblivious to all but his heartbeat. A infinite time later he heard a voice asking him to slow down. He smiled. The sitar player had sat down beside him. Maybe his pounding was music after all.

20

Listen to your heart beat,
let your fingers dance,
hum along.
The pain is not so bad.
Hands entwined in his mother's hair.

<u>Singing in the Rain</u>

infinite
spherical
loops.
This could go on for ever.

Fuck it! I'm goin' to stick it right here. I've used a bunch of editors to do this thing and it looks different in every one: Some where I say some thing about Einstein comin' up with $E = mc^2$. Turns out it didn't originate from Einstein. Some guy by the name of Friedrich Hasen(suppose to be two dots above the followin' letter)ohrl was usin' it before him. I've thumbed through DISCOVER PRESENTS EINSTEIN

Turns out Einstein made a bunch of mistakes. Don't we all? And like the magazine says: Most physicists would be proud to have made the kinda mistakes Einstein did. What a man.

IV

loop

He had trouble defining himself, so he let others do it. Or at least, he tried. Their definition never quite fit. One psychiatrist had diagnosed him as having schizophrenic affect. The psychiatrist's explanation: A loner with peculiar thoughts who had done a lot of drugs.

First thought: everything is relative. No, he had not done a lot of drugs. Not compared to the crowd he had run with. Not compared to what he could have done. He had actually only bought one ounce of grass in his life and his room-mate's dog ate most of that. High at the time, he didn't care. He had turned down offers to smoke for a long time. He had been accused of being stoned even when straight, so why bother?

Infinite

spherical

loops.

Memories of babbling monologues, Dunkin' Donuts poofing the donuts being one. (Hada sorta Jonathan Winters, Robin Williams thing going on at the time. The people he was with then found it kinda amusing. They don't anymore, so he doesn't do that kinda thing anymore.) He had gotten to third base by accident in a Dunkin' Donuts parking lot. He had been feeling her breasts, small but firm. His hand relaxed, her jeans were loose, he found himself in unknown territory. Scared, his hand jerked back out. She pushed it back in. What the hell? Relax. Enjoy. Her panties were silky, her privates moist. Very interesting. Curiosity won out.

The Doors said it first: People Are Strange.

Another trip back to the psychiatrist, this time curious to find out which thoughts the psychiatrist found to peculiar. He hadn't shared that many with him. He was there only for a 15 minute Med Check after all. And many a time much of that time had been spent waiting, patiently, for the psychiatrist to finish typing on his computer. Maybe the psychiatrist found that to be peculiar? There was a time when the sessions had been longer, the psychiatrist less busy. He had shared a thought back then. Had to do with computers, the human mind, and ambiguity. The human mind can deal with ambiguity much easier than a computer can. Some thought. Well, the basic component of a computer is a switch. On/Off. Yes/No. Much trouble is gone through to make certain that a circuit can only yield a Yes/No answer. The switch in the in the human brain? The Synaptic Gap, the space between nerve endings. This gap is bridged by neurotransmitters, chemicals.

There is a variety of neurotransmitters, their effect on the emotional thought process is well documented. The threshold at which the nerves actually fire varies. Eureka! Infinite Variability. Maybe, this is how the human mind deals with ambiguity.

Tries it out in a research class, wants to see what others think. Starts out with a clip from The Island of Lost Souls. Wishes he had Devo's Are We Not Men. Darwin in many forms. Are we not men! These are the laws!

22

Loop

Signs, signs, everywhere there's signs blockin' up the scenery, wasting my time. Do this, don't do that. Can't you read the signs.

Sign on a grocery store door: No shirt, No shoes, No socks, No service.

No shirt, No shoes, No problem. But socks. Just make sure your pants are long enough to cover. Won't work in the summer when you wear shorts though.

Loop

(He stops typing for a while. Saliva in the background. He pounds his knees.)
(Got to stop that foot from tapping. Keep typing. Turn up the stereo!)

Points out that he is the only native English speaker in the class. A gasp from the crowd! Just an observation. No offense meant. No offense taken. He ignores it. Unfamiliar pronunciations, intonations, rhythms. Somehow, he makes sense of it all. What a surprise.
To make a long story short. Only three people left at the end. Only one has a question and he thinks you're crazy! Probably am. Sounded good to him, though.

Hands in his mother's hair.
Infinite
Spherical
Loops.
Pain Be Gone!
Not quite yet.

Does any of this make sense?

This could go on for ever.

V

loop

(If he can find that paper entitled <u>Random Order and Infinite Variability</u>, it goes here.)

(He has made one attempt to find that paper so far, unsuccessfully. Probably won't find it. He has found some things he wrote in what he referred to as scrap-paper. He's been known to write notes to himself, make diagrams of things he was working on. He's into this reuse/recycle thing. Just because a paper has writing on one side doesn't mean he can't use the other the other side for notes. So he's saved a lot of things for that purpose only. The teacher did write a lot of comments on the back, so he probably thought: I can't write on the back of this, might as well throw it away. That leads to a couple of things. He usually doesn't bother to go get that scrape paper he's saved, just grabs what's ever nearest; the back of an envelope, paper that already has writing on it that he just writes over. And he does have a vague memory of what he wrote in that paper. Had to do with: Well, may be the Universe is trying all these different things and what we see, what we pick out are the things that make sense to us. That thing about monkeys writing Shakespear in the beginning. What he did there was paraphrase an example used in probability. The original goes like this: if you take enough monkeys, put them at enough typewriters, give them enough time, one of them will eventually write a work of Shakespeare's. He remembers using that example in the paper. And that thing about The Observer alters the Observed which... That he derived from a statement used in Quantum Physics. If you don't know anything about Quantum Physics, it might take a while to explain. And the fact is, he doesn't know that much about Quantum Physics. Just a few basic concepts, ideas. And even if you do know something about Quantum Physics, you may not recognize that is where it comes from. He's played around with it a couple of times, turned it over in his head, and that's what he came up with. The simpler he can make something, the more likely he is to remember it. There is probably a better way of saying it, but for now, that will have to do. He was trying to do laundry at his Dad's, had to come home to get something, had been thinking about a couple of things, decided to type them down. There is more he can say but he really needs to get that laundry done. And he's not even certain he can save what he just typed. Something screwy is happening to Word Perfect and Windows. Something about no backup file. He'll have to figure that out later. He really doesn't want to do this all over again. He has written some of it out ahead of time, but he made changes when he typed it in. Sometimes he types about what is going on while he's typing. So he's not going to remember exactly what he typed before. He should save things more often.)

VI

loop

(Frank Zappa in the background singing: Don't cum in me, don't cum in me. Originally, they had to record that phrase backwards, but things have changed. He kinda liked the original better. Made you think: What did he just say?)

Hands entwined in his mother's hair.

(Frank Zappa again: There will come a time when you can even take your clothes off when you dance. Making it kinda hard to concentrate on typing. May be he should turn it off.)

Infinite
Spherical
loops.

Just another brick in the wall.

(THis is where
$E = MC^2$ goes.)
Another trip to the psychiatrist, this time he points out that he was a loner because he expected others to find his thoughts peculiar. Had a discussion with a co-worker about evolution; to him, that's all it was, a friendly discussion. The co-worker treated him as the devil-incarnate afterwards. Seems to be a conflict. The existence of God disproves evolution. Evolution disproves the existence of God. He doesn't understand. The simplest definition of evolution: things change. He sees that everyday, everywhere. Einstein said that God would not simply roll the dice. He asks: Why not? What is wrong with a God curious to see how things turn out, has faith that things will turn out. It is said that we are made in His image. We have curiosity, faith. Is that too simple? $E = MC$ squared. A variable equals a variable times a constant. One of the simplest equations in math. The genius of Einstein's was not that he defined this relationship between energy and mass, others had probably done so before him. The genius was that he did it so simply. A variable equals a variable times a constant. Awe. What a surprise.

Infinite

spherical

loops.
This could go on for ever.

25

VII

infinite

Hands entwined

spherical

in his mother's hair.

loops.

His mother didn't always understand why he did what he did. That's alright, he didn't always understand why he did what he did either. But most of the time (harmonica in the background), she seemed to accept him.

loop

He has often been asked if he hears voices telling him what to do. He says: No. But he does hear voices telling him what to do. It's just that they come from real people.

loop

Most of the time. There was an incident. He had been unemployed for a while, watching TV, sleeping on the couch. Memories of haircuts were hot, humid Indiana summers, neck itching. He had let his hair grow long. Nicks, scrapes. (Bill Cosby said it first: Zip, Zap, Your face is cut to shreds.) That was his memories of shaving. It was a everyday ritual he didn't need. It was the early 70's. There was a subculture that accepted this, but it was, after all, just a subculture. He had become discouraged.
His mother got on his case. Slap across her face. Sister astonished. He got on his motorcycle (350 Honda), heading off to see a woman he had been dating. Nice small breasts. Firm. Passed a car on a hill. Headlights in his eyes, he swerves back to the right. It wasn't till later, feeling those breasts, that he realized just how close those headlights had been.

He shakes.

infinite
Hands entwined
spherical
in his mother's hair.
He hopes she forgives him.

This could go on for ever.

26

(Just scrollin' through, seein' what Ive got, This page was empty, so I thought I put somethin' here.)

I'm goin' put somethin' else here, then see if I can find Pink Floyd's "The Dark Side Of The Moon". (Found: My Spirit Flies To You" instead)

It's Tuesday, 3/31/2009, 1:00 P.M.

I just come home from havin' lunch with Dad.

He put his hand on my arm.

And left it there for awhile.

I think that's the first time he's done that.

At least in a long while.

(and I'm rockin' back and forth
tryin' to keep from
cryin'.

It's not workin'!)

(Found some Kleenex.)

(But I'm still just goin' to sit here and rock for awhile.)

On With the show!

27

VIII

loop

Call any vegetable, and the chances are good, that the vegetable will respond to you. Frank Zappa, again. Another strange M/F. His cousin turned him on to The Mothers of Invention and Steppenwolf, too, while he still had a flat-top. He especially liked (sing along if you want to):

What's the ugliest part of your body?
What's the ugliest part of your body?
Some say your toes.
Some say your nose.
I think it's your mind.
Your mind, Your mind.
Woo, Woo.

(Hello Frank Zappa.)
(I'm channelin' now)
Too bad Frank's dead now.
I really woulda liked to have met
the guy. And thanked him.

Let your fingers dance.

Strange relationship with his cousin. At times they were like brothers. Then there were long periods where they would not see each other. Now was one, a very long period. His cousin was what would now be called a player. He wasn't. His cousin smoked dope. He didn't. His cousin did expose him to alcohol, though. Road trip to South Bend. Earlier, he had seen Bill Cosby there, at Notre Dame. Hamm's. Warm. He didn't much care for it at the time, that would change in the future.

Really change.

God Damn The Pusher.   I'm not refferrin' to you, Tim.

Several directions he could go in now.   So many paths to take.

One.

A friend tried to teach him how to play The Pusher, (Hoyt Axton wrote it), on the guitar while he was in San Francisco on that quest to be a rich-and-famous movie director. He didn't have much confidence in his left hand. Rhythm with his right, no problem. But his left? Not sure how close he came to playing it. (Thanks for tryin', Rick. And if Scott is still around, tell him hi for me, would ya?)

Two.

Time trials at the Indy 500, several years later. Late 80's. The 70's had been wild. Woman exposing their breasts. Not much of that now. A case of beer, Old Milwaukee this time, all that

he could afford. Leaving the track, then, out the windshield of his 87 Nissan Hardbody that he bought,
brand-new, in San Francisco when the Datsun 1200 coupe he had driven out there in, loaded to the windows with crap, pooped out:

Strange
A woman running toward him.

Strange. Something wasn't quite right.
She was on a right angle to the windshield.

He realizes the truck is on it's driver's side. He smiles, rolls down the passenger side window, undoes his seatbelt, felt uncomfortable without one, and climbs out. Sits there on the edge of the roof for a while. Ambulance, Police. Yes, he had been drinking. Wasn't a very good liar. Several witnesses testify that a car pulled out in front of him. He swerved to avoid hitting it. He doesn't know. Blackout. Not his first, but his last. Not going to get away with it anymore.

Knows he'll be court-ordered to undergo counseling, so he beats them to the punch. Goes in himself. Not the first time after all. The therapist literally condemns him to hell. Literally. He'll never stop drinking unless he admits to being an alcoholic. That definition thing again. He says he doesn't know. Not that he isn't, not that he is. He just doesn't know. Sure, he had some of the traits of an alcoholic, but self-medicating seemed more apt. He was more of a binge drinker. Didn't drink all of the time, didn't get drunk every time he drank; but, to get that wild hair out of his ass, when he got drunk he got very, very drunk.

Hands entwined in his mother's hair.

Infinite
spherical
loops.

Sometimes, all I can come up with is one word.

The therapist writes a nasty note to his lawyer. The lawyer asks him about it. Catch-22 was all he said.

loops.
spherical
infinite

Hands entwined in his mother's hair.

This could go on for ever.

29

VIIII

loop

Sometimes light acts like a particle, sometimes a wave. Which is it? Why not both. A wave is
just a way of describing a series of single points after all.

("Take it easy, baby. Take it as It comes". Jim Morrison again. Believe it or not, All these references to music came
about because that's what i was listenin' to. I'm tryin to write, concentratin'
on that.

and all of a sudden. A lyric pops in to my mind. And I go: gotta write that down.
("All the children are insane".)
(Jimmy again.)

(I'm back! Got another cup of coffee. this might take awhile.)

(Got my pipe.)

(Took a shit.)(where was I?)(LOL!)

(I really do like writin' this way.)

(Anyhow, It's not until after I've written it down and read what I've written , that I go: That kinda fits here. I really am
lettin' my sub-conscience be my guide on this.)

(And do you know what I like about stickin' comments in like this. It's kinda how I operate.)
("Take another little piece of my heart if it makes you feel good." Janis this time.)

("Don't you cry." Thank you Janis.)   (And the medical system. About alot of things.)
(Boy, I've got a lot to say about the psychiatrict system. But not right now.)   (I gotta finish
this
(I'm always writin' notes on things.)   damn thing!)

(Again, that sub-conscience, zone thing goin' on. this wasn't what I planned to write here.)

30

X

loop

On the way to work, Eddie Vetter singing: I'm still alive.

What a surprise!

He had an IQ measured in the genius range. Notice he says not that he was a genius, just an IQ in the genius range. In the hallway at lunch in high school: A trouble-maker justifying the trouble he makes by proclaiming himself to be a genius. A teacher points out: genius is not a level of intelligence, it's what you do with the intelligence you have. He hadn't done that much with his intelligence yet and I'm still wonderin' about that. I still haven't come up with a better explanation for the Observer... thingy.

He scored highest in language. What a surprise! He often was tongue-tied when he tried to talk. He usually sucked on Vitamin C drops at work to keep his mouth moist. They often popped out when he talked. At a Mensa conference on consciousness, he tries to ask the speaker a question.
Rambles on, something about digital vs. analog, totally off point as to what the speaker had talked about, loses his train of thought, so many to chose from, and says so out load. The crowd laughs. He smiles. Nothing wrong with playing the fool as long as people laugh. It's when they grow concerned... He gives a woman he met there a hug before leaving. Melt-in-your-arms soft.
Married though.

Some one said it before: Love having written, Hate writing. A teacher, film genres, once described his writing as terse. The teacher wasn't sure if that was good or bad. He wasn't either. He didn't follow the rules of grammar. Dropped pronouns. Used incomplete sentences. Run-on sentences. Dangled participials.(He has some idea as to why he might dangle participials now. He's usually referring to something in the sentence before and it seems rather repetitious to repeat it.) Probably made it hard for others to follow. Sorry about that. It's just the way he thinks. Same teacher, now a published author himself. Official Policy: A letter grade off for everyday late. He never turned in a paper less than a week late. It could take days, searching for the right word, the right phrasing. Still got B's, C's. Don't know if that said more about the teacher or himself. The teacher even suggested he submit some for publication. Never did. Sorry about that Wes. It's not that he wasn't listening. Just, at the time, he wanted to make movies, not write about them. Didn't realize that writing might lead to making.

Infinite

spherical
loops.

Hands entwined in his mother's hair.

Is the pain still there?

A little.

This could go on for ever.

You know, I started out bein' pretty orginized. But as is my habit, things are gettin' out of control.

XI

LOOK WHAT THE'VE DONE TO MY STORY, MA

Mike Barden

INF-101 51

I found it while browsing through a used bookstore. There had been a fair amount of publicity surrounding a film being made from it, particularly about the movie's star, 1st time actor, David Bowie. I had a extra buck, so what-the-hell. I bought it. I'm glad I did. THE MAN WHO FELL TO EARTH by Walter Tevis (also author of THE HUSTLER) was most enjoyable.

But the joy I felt while reading the book was not there while viewing the movie. I began to wonder if the filmmaker, Nicolas Roegs, and I had read the same book; or, had he read the book? I was numb from disappointment. The movie had missed the point.

The original story is short, just 144 pages. Much of the dialogue takes place in the character's minds, something easily handled in written words, but which can cause problems for film. But it is these interworkings that are the heart of the characters, the drive. It is up to the filmmaker to unearth the heart. When they don't, when changes are made that trivialize, you wonder why.

Filmmaking is an egocentric endeavor. But too much ego can distract from the story. Roege's hand shows early on. The scene of the crash of the spacecraft in quick repetition. Is he afraid that I will miss the point if only shown it once? Or is he just so enamored with the image. The fact the spacecraft sinks into a lake in the movie is also a major deviation in storyline. Its use as a tourist attraction by the farmer who finds it is a major factor in the downfall of Mr. Newton, the original owner, in the book.

Excessive Ego is often justified by making a "artsy" film. Newton's reminiscences of his other-world homelife were artfully filmed. But strange suits and strange vehicles in otherwise everyday Earth-type activities do not make an alien, even when shot in silence. The years of preparation for infiltration into Earth's society formed the memories of Newton in the book. The deficiencies of this preparation led to Newton's sorrow. After all, it was based on information gleamed from TV signals wandering through space. How much of our own concepts of this world are based on TV signals. Amazing Discoveries, our interplanetary ambassador.

Sex sells. It is, after all, the lowest common denominator of mankind. Its use can increase a movie's popularity. But increasing popularity at the risk of demeaning content? Tevis' subject was alienation and loneliness. Roegs' subject was an alien and sexual frustration. More accessible, but not as satisfying.

Newton was peculiar; generous and gentle, but slightly incomprehensible. Those around him could sense his alienness, but were drawn to him by his humanity. David Bowie had an air of other-worldliness. His physical attributes, his ambivalence, contributated to the alienness. Missing was the humanity. Newton knew how to smile, how to hide behind a smile. Bowie never smiled.

A pervert bent on kinky sex with a beatiful coed is how Dr. Bryce is introduced in the film. Interest in the film the coed uses to record the tryst, leads to Newton. Tevis' Dr. Bryce is an introvert, immersing himself in work and movies after his wife's death. His disgust with the academic system and a mild case of hopeless despair, a feeling that he is going nowhere; worse yet, that there is nowhere to go, enables him to follow his intrigue with the image on the screen to Newton. Perhaps sensing their commonality, Newton confesses his origins to Dr. Bryce. This confession is overheard by the CIA, who's keeping tabs on the man with no past and marvelous inventions. (They say necessity is the mother of invention.)

Newton's other confident is Betty Jo; in the movie, a beautiful, hillbillish, young woman. She is made aware of Newton's origins during a bizarre and frightening sexual escapade. Candy Clark

(Hello Frank! This time it's desperation.)

is a beautiful woman with a fine body, and her nude scenes warmed this voyeur's heart. But again, the substitution of the obvious for the sublime. In Tevis' world, Betty Jo was a woman who found her niche in the American system through the welfare department. Pleasant looking with a few extra pounds not worth the effort to get rid of. A middle-aged child living for her sugar and gin. Newton's preconceived image of Earth, shaped by TV, was of a comfortable middle-class, constantly striving for more. Here was Betty Jo, with less, but content. Betty Jo loved Newton, the love one feels for a stray cat. And surprisingly, Newton, from a race not known for its passion, cared for her. But sex was too alien to consider, doomed by incompatibility. The poignancy of what can never be, as opposed to what has failed.

To realize that it can never be. Newton's people were to leave a dying planet, immigrating to Earth. In exchange for a new life, they would give Earth, seeming determined to self-destruct, a chance at life. Surreptitiously infiltrating into positions of power, they would this world to a safer course, their right to do so a source of discussion between Newton and Bryce. By the time Bryce concluded that it may be our only chance at survival, it was too late. Newton was a broken man. Both purposely and inadvertently, the government had thwarted his plans. But more damaging, in living with the apes, Newton had become lost in limbo. No longer an alien, he could never be human. His race could never escape our influence here on Earth. Their best laid plans would go astray in a sea of sugar and gin.

Plans gone astray. Who is not effected? By looking through another's eyes, we often can better see ourselves. Had Roeges made a commitment to look through Tevis' eyes, He might have had a wonderous thing.

XII

loop

Walter Tevis. Another strange M/F. Read three of Tevis' books, all about alienation. His understanding, though, is that he was a she.

(While reading this review again, he thought: Boy, this is pretty negative.

(You know, he's thought about this some more and he came up with this: Here's David Bowie. He's kinda famous and I criticize him. Nicolas Roegs. He's kinda famous and I call him an egocentric. And here I am. A student at a obscure school. The irony of it all! Maybe he's starting to regain that sense of irony.)

Thought about that and he's come up with two things so far:

One.

May be if he hadn't read the book first he would have liked the movie. But he did read the book first and it did seem to trivialize what he experienced while reading the book.

Two.

When he was enjoying a movie it was because it was just clicking, kinda made sense, he wasn't thinking about what was going on, more experiencing what was going on. It was what bothered him, jarred him out of being in the moment, that he really noticed and thought about: Why did that bother me so? If he had written about the movies, it probably would have been mostly negative, and he wasn't certain if that would have been such a good thing.

While typing this whole thing in, he come up with a couple of more thoughts:

One.

The teacher kept bugging him about getting his papers in on time. Said he'd get A's if he did. The teacher probably wondered why he never did. Again, sorry about that Wes. His writing was kinda informal, breezy, and it might have looked like he knocked it out in one night. He knew big words, but he didn't necessarily think that way. He was no Bill Buckley. He liked to keep things simple when he thought, and didn't have much reason to use big words when he talked. The people he was often around didn't use them, and seemed confused when he did. So, again, it could take days, searching for that right word. And finding that right word seemed more important than getting an A.

Two.

36

He was a lousy typist, still is. He had his mother type up his papers. She was very religious. Cuss words, a lot of talk about sex. He wonders what she was thinking when she typed this paper up for him.

Hands entwined in his mother's hair.

Again, he hopes she forgives him.

Infinite
Spherical
loops.

This could go on for ever.

(His Dad just called. He didn't finish his laundry the other day, so his dad did. Thanks Dad.)

Three.(a quite a bit later.)

Like he just said, he'd write his stuff out on paper, have his mother type it out for him on a typewriter. It was the early 80's. Home computers weren't that common. And he didn't like them that much anyhow. Still don't. There should be an easier way of doin' this! So, when he typed that paper into WordPerfect, spell-checker underlined wonderous. He thought: Maybe I mis-spelled it. He had already noticed that they've in the title was mis-spelled and decided: I'm trying to report things as accurately as I can, so if it's mis-spelled in the original I'll just leave it that way. But he got curious. So, he took a look. And there were a couple of suggestions. Then he thought: Maybe I made that word up. Maybe, I should change it. Wonderful would do. But you know, sometimes it's not so much about what a word means as to how the word sounds in his head. (Kinda has a ring to it, kind of sounds flat) And he kinda likes the way that word sounds. Most people know what the word wonder means. They have some kinda idea what adding -ous to a word does to it. So they should get a feel for what he means there. And he wouldn't be the first person to make a word up. It happens all the time. They keep havin' to up-date dictionaries. So, he likes the way that word sounds and he'll leave it alone.

(I'm reading it again and I've found a bunch more of mistakes. LoL!)

(So that's what you meant by cleaning it up, Wes!)

(I hadn't read the typed version before handing it in.)

LOL

37

We'll throw this story in now because it kinda fits.

When he was in grade school, he went out for the school choir because his friend did.

The teacher doing the auditions said:

You have perfect pitch.

He didn't know what that meant then.

Wasn't sure if it was

a good thing

or

a bad thing.

He does now.

But he doesn't sing out loud much.

He just hums.

If others hear him, they'll probably think it's weird.

Which leads to another story:

He was at his nephews helpin' to fix the place up.

His Dad says:

Is the radio on?

He says:

It's just me humming. When others hear me, they think it's strange.

Dad says:

You're just self-conscious.

Dad, I didn't even realize I was hummin' till you said somethin'.

(Another empty page! I'lll think of somethin' later and stick it here!)

(one more point I want to make: It's not that I blame people for treatin' me like a smilin' imbecile.

I am at times.

Just not all the time. 3/31/2009 1:30 P.M.)

If I haven't lost you yet,

Here's the deal:

I really am trying to write a book entitled: Welcome To My World: Come on in. The Water's are fine.

But I've never tried to write a book.

Small papers,

Yeah, but not a book.

There's been a lot of serrendipity, happenstance, coincidence,

Involved in this endeavor.

And I went with it.

Sometimes, I wonder if I should of.

And each time I think I've gone too far

I go a step further. WIP applies to me, also.

Forget the stuff written in this font.

I'm just playin' around,

seein' what I can do. Not that, that ain't happenin' in the
Larger type, too.

(One problem with this editor: I don't think it's got SpellChecker.)

Concentrate on the other stuff written in the larger type

Let me know what you think.

Comments can be sent to mikebarden@gmail.com

Here's the kind of things I'd like to know:

I'm goin' too far, is it too hard to follow. I lke the way things are happening but I want other people to be able to read it.

I've started a bunch of things, some I do plan to continue. You got any in mind?

There are other things I could talk about. Something new happens all the time. Memories keep poppin' up.

Which leads to the second deal, but I'm close to the bottom of the page so I'll find some other place to stick it.

(Believe it or not. "Wicked World" by Black Sabbath was playin' >>>>) And it's about time to go work. Bye for now.

XIII

He had some other things he had written about ahead of time that he was going to put here, but may be now's a good time to say a few words about his Dad. He's thinks he's all ready written about how his mother seemed to accept him. Like he said, he's changed a few things around, not sure where things are any more. He's never been certain that his Dad completely accepted him. His Dad has said he just wants him to be happy. But for his Dad to see him as happy, it often seemed like he had to be interested in what his Dad was interested in, do things the way his Dad did them. And that just wasn't his way. He had tried. He had been working at this last job for 9 years, his previous record was three and a half. Then this pain thing came up. His Dad has said he respected him, had learned some things from him; which, at the time seemed kinda weird: I'm suppose to learn from you, Dad, not you from me. And he respected his Dad. Found out his Dad had wanted to be a high school basketball coach, but had to drop out of high school to help support the family. His Dad was usually there when he wanted to play catch, but his Dad never forced sports on him. So, he respected him for that. Fact is, his parents never forced much on him. Sure, they tried to guide him, canjol him into doing things. When he quit going to church, they'd often say, on their return, so-and-so missed you. But they rarely said: Do this or else. Some may see this as permissive. To him, it was more progressive, which if you knew his parents, as straight-laced and religious as they were, seems kinda strange.
And things had gotten better between his Dad and him. Toward the end of his time off from work he had gone mad, lost touch with reality, completely bonkers; for a few days, at least, clinically insane. At least once, he did see something he now knew was not there. And one time, after coming home from his Dad's, he kept looking out his window to see if his truck was there. It wasn't. Thought: Maybe Dad brought me home. Decided: No, I drove home. So, he stared out the window, and, very slowly, the truck appeared. So, he had, for a while at least, lost touch with reality. And, this may seem strange, the idea that he had lost touch with reality for a while didn't bother him that much. His behavior, in one incident at least, apparently frightened some people, and for that, he apologizes. And another time, coming home from a rather bizarre road trip to Indianoplis, he kinda blacked out. Found himself weaving on the shoulder of I-69 in heavy traffic. And he has thought: could of hurt myself or some else then. Again, he apologizes to any other driver on that road at that time.
But for the most part, his behavior was fairly benign. Wasn't certain if he had that good of grip on reality in the first place. At least, his reality often seemed different from those around him. It led to this writing. He had a creative side that he had thought he had to stifle to live the life he was living. Now, he saw he couldn't. And during this period, he became a little more loving, opened up some to his Dad. Said some things he might not have said otherwise. Now that he was back to the everyday grind of work and chores, he was starting to close up again some. Get irritated at his Dad for worrying so much about him. But maybe he could stay just a little bit insane.

Back to the show!

Hands entwined in his mother's hair.

40

If nothing else, they say what don't kill ya , makes ya stronger. It hadn't killed him.

Infinite

Not this time, at least.

Spherical
Loops.

This could go on for ever.

OK, this is blank space, so I'll stick this here:

Two things first then I'll give you the second deal.

First,

A big thanks to Dr. Paul B., a computer science teacher I had at B.S.U. He treated me as kinda an equal. Even though I really wasn't comfortable with that. He asked me what I knew about the theory of relativity. I said: Nothin' He suggested I read it. I had been scared to. Thought it probably had all these mathematical equations that maybe I woulda understood in high school but had lost my feel for.
But since he thought I might have known somethin' about it, I decided to give it a try. Found it in a used bookstore.
And thank you Einstein.
For puttin' it in such simple terms that even an smilin' imbecile like me could understand it.
Startin' to tear up again.   You know, it really had been a long time since I'd cried. Not anymore.

A little Saliva now. get me goin' again. "Wecome to the show." & "Please come inside"

&

"Ladies and Gentleman"

LOL

"Turn up the stereo"

I got it as loud as it will go. LOL

Second: then I gotta go eat lunch with Dad. finish the rest later.

Told a co-worker that I thought I could poke a hole in the laws of probability. He looked at me strange.
I really think I can.
I really want to put it in the other book and give a longer explanation why I think that.
But for, now: Go to About.com>Mathematics>Philosophy of Math>Questioning A Basic Principal of Statstic(Wouldn't let me put in the final s)
Forget the journey of... part. If I wasn't crazy by then, I was headin' that way.
But now that I'm theoritacally back to normal(Ireally wish this damn thing had SpellChecker. Anyhow.)
I now know why I thought a infinte set was just a series of finite sets.
It was how you were shown to construct one in a textbook I once had. Don't have that textbook anymore, so I'm not 100% certain. Rarely am.
If you can show me my error.
Please post it there.   (maybe this explanation is long enough, and there might not be a second book anyhow.)
                              (I'll put this in here before i go to lunch. A fuller explanation involves this experts vs. common folk thing I got goin' on.)

Before I went to lunch with Dad, it was my intention to put the second deal here. I see I'm close to the bottom of the page and that wouldn't have worked anyway. So, let me see how much I can get in here. I'm gettin' disorginized so I don't know if I've mentioned this yet, but I've opened up mpre with Dad, Talkin' more, and I'm findin' out we have more in common then I thought.   Go to next page towards the bottom.

41

XIIII

Loop, loop, damn loop.

What to talk about next? No children that he knows of. Can't talk about that.

There was this time in the San Bernardo Mountains. Early 70's.

(The pain returns. Who will believe this? Gotta get rid of the pain. Keep going.)
(Growl)

He spent a lot of the early 70's on the road, seeing the U. S. Of A., when things got too much for him. Living in the back of a Datsun pick-up. The simple life. Things aren't that simple now.

(Become a child again. Keep that sense of awe.)

Arrow Creek. It was in his guide to nude recreation. He liked getting naked. Not always that way. He wasn't circumcised. Made him different from the other little boys. Use to hide behind shower poles. Not any more! That time in Arcosanti, getting into the zone. Stood, naked, on a mound of earth during a thunderstorm, daring the lightning to strike. If others saw him, they probably it was weird. But it felt good.

Hands entwined in his mother's hair.
Infinite
Spherical
Loops.

Anyhow, my mind is kinda racin' right now, light bulbs goin' off all over the place, many paths to take. I'll try to stick to just one. Found it! This was on paper. I'll start with it.

Delusions = Distorted Thinkin'(I got a few of these.)

I like science fiction.

I like to dream of the future.

But I'm a contridiction.

A dilema.

I also kinda like the way things

were when I was younger.

There's a word for that.

But I keep thinkin'

Anarchistic. Go to Page 44

42

XV

(Memories can be bitter-sweet.)

loop

The woman he should of married. The woman he could of married. The woman he stalked?

The woman he should of married. Nancy. She first showed him the infinite. Small, firm breasts. Her ass wiggled even when barefoot. Awe. Fell asleep once while she tried to explain the experience. Sorry about that. He was exhausted. Wanted to dream.

Hands entwined in her hair.

The woman he could of married. Talked about it. She was German, Lithuanian. Good at languages. Her English had a lisp that he found enchanting when they first met. A month later, it was gone. Damn! Small breasts. Not so firm, but he was older then, too. European, a lot of body/pubic hair. Nothing wrong with that.

Infinite

They met at that clothing-optional, hot springs resort in Northern California. She was looking for a New-Age community to live in. He was getting ready to head back to the real world.

(Unemployment running out. Go live with parents.)

She was wanting to check out one in Southern Indiana. He was going back to Indiana. Want a ride?

Spherical

The community in Southern Indiana was a bust. End-of-the-world religious bunch. He took her down there. They talked to him rather to her.

She wanted to work in their cabinet-making shop. Women could only teach.

Ramona. Spanish first name to hide the fact that she was part German, Germans not being particularly popular in Lithuania at the time. Left-over from WWII. Her father had been a cook in the German Army. Met her mother. Deserted.

Only was with her for six months. Much of that time she was elsewhere, looking for her place in the world. But he knew more about her than any woman before or since. How strange.

Loops.

Before she left for New Hampshire (New-Age community there) had a meal of sliders from White Castle (all he could afford) and white wine (Given to him in San Francisco). Not bad, she enjoyed it.

This could go on for ever.

(The woman he stalked? Not ready to go there yet.)

And I know that's not the word.

I like William Gibson.

The system has broken

And we gotta start all over again.

Don't cry.

LOL

You know, I am a bit of a non-conformist. I do break rules sometimes.
Some times it's because I didn't realize there was a rule against what I did.
I actually can be a bit naive. And when I realize I fucked up, I often fuck up
Again tryin' to fix my first fuck up & .... That happened with the publisher. Which
is why I thought they might be pissed at me. But I got another e-mail from them
that was nice.
Sometimes I know there's rule against what I'm goin' to do but I just gotta break it.
I gotta do what I gotta do. I'm willin' to suffer the consequences.
And I don't put many rules on myself, but sometimes I even break them.
I do that below. I don't like to be preached to, so I don't like to preach.
And here I am preachin'!
So let's put it this way: This works for me. Maybe it won't work for you.
But can you at least try it?

What me and Dad have in common: We both kinda like the way things were better.

The next step in the process. "Turn up the stereo" Didn't want to take the time to find another Cd so I'm re-playin' Saliva.

Awhile back a supervisor of mine subscribed to a music download service. Didn't want them to take music offa his computer,
knew I knew somethin' about computers. So he asked me how to do it. He probably wanted to know about firewalls. I know nothin'
about firewalls. Don't have to. I'm only on the internet when I need to be> I'm off it when I get done> Don't really have anything on my computer worth
stealin' anyway. So, I know nothin' about firewalls. Just that they exist.

So I said: get off the internet and turn your computer off. That's what I do.

It's April 1st and there's all this worry about a virius. I'm not worried. I don't download that much. And if it's a worm instead of a virius, I'm not on
the internet that much

Sorry about this, folks: There is only one way to be sure someone else won't get into your computer.

Get off the internet and turn your computer off. You're just wastin' electricity by keepin' it on when you're not usin' it.
Sure it takes a while for it to boot up but it's cheaper to turn it off.
And as long as you're connected to the internet, someone else can get in it. Firewall or not.
The reason they keep updatin' firewalls. Someone found away to get around it.
You're just addin' expense to expense to expense. I really wish this thing had SpellChecker.

Anyhow, that does kinda lead to the second deal:
If there's a subject I haven't touched upon and you want to know what I think: Put it in a e-mail to me. If I can come up with a story or even an
opinion, I'll e-mail you back, I'll put it in the e-mail so you don't havta worry about a virius. and if you like it, let me know and I'll put it in the book.
Put your name with it. Might not just make you rich, but might make you famous also. How's that for a deal.

And don't ask about firewalls. I know nothin' about them. There's a lot I don't know. And there's always someone   (Page 47)

44

XVI

loop

The first alcoholic blackout occurred in New Orleans during Mardi Gras. Early 70's. Woman exposing their breasts. Drinking half-pints of whiskey in quarts of beer. Looks down at his drink, thinks: I'm not going to be able to finish this. Next, he's curious as to where he is at. Can't read the street sign, tries to climb it. Can't. Realizes he doesn't have his glasses on. Checks his pockets. There, in the back pocket. His glasses are mangled. They're wire rims, he straightens them back out. Doesn't recognize the street name.

A cop pulls up.

Do you know where you're going?

Farmer's Market.

That's several miles in that direction.

Thanks. (It always helps to say thanks to a cop.)

Next day, he awakes. His chin is sore. Climbs out from the camper on the back of his truck. In the cab, looking in the rear-view mirror. Rash on his chin, front tooth chipped. Maybe, that tooth would look good in gold.
That tooth is still chipped. A dentist has tried to fix it several times. He smokes a pipe, chews on it. The tooth keeps breaking. Maybe he'll try gold next time.

Hands entwined in his mother's hair.

Infinite
spherical
Loops.

This could go on for ever.

(I bit down on my pipe again. Now the whole damn tooth is missin'. I've got another tooth that broke off right at the gum line but it don't hurt yet. I need to see a dentist but my old dentist isn't under my new insuarance!)

("It's all the same fuckin' thing, man." Janis again. It's like she's talkin' right to me.)

Don't cry!

(I got another story to tell about that tooth.)

(Just got a Kleenex to blow my nose.)

45

XVII

loop

Sometimes he feels like he is misinterpreted. People put a different meaning on what he said than the meaning he meant. Has thought about the accident with the truck. Some may think: he was drunk and he avoided hitting a car. He's condoning drunk-driving.
Here's the way he looks at it: I totaled out that truck. It was a nice truck. Maybe if I hadn't been drunk, I could of avoided hitting that car without totaling out that truck.

(walkin' in circles, again. Thinkin' about callin' Dad and tellin' him I'm not comin'
to lunch.)

(I don't think he eats if I'm not there.)

(I'm goin' to lunch, if it takes longer to do this damn thing, so be it!)

but I'll put this in first

i quit drinkin' for a long time. But I got a taste for Guiness. So, I started again. Remembered I liked brandy.

So, I bought some of that.

Remenbered   (Goin' to leave this mis-spellin' alone so you no that I make mistakes.)

I kinda liked gettin' drunk.

it helped me get that zone thing goin'.

So, I got drunk one night.

on purpose. first time for that.

(Tool: where's that damn Cd jacket!

"Trust me."

Couldn't find the jacket.)   forgot that I was suppose to eat dinner with my parents,

My mom was still alive.

My sisters showed up at my house all worried.

And I wasn't even that drunk!

46

XVIII

not for me, at least.   haven't drank since.

loop

The early 70's were a busy time for him. School, road trips, work. First went to college at a small Christian school. Wasn't particularly religious by that time. Saw Jesus as a good role-model, but not certain about the supernatural aspect. If it took being supernatural to do what Jesus did: be kind, loving, accepting, generous, forgiving, patient, all those little things Jesus did so well; Well, what chance did he have? He wasn't supernatural! And even Jesus lost his temper once, at the moneychangers! So, he wondered about that.

Hands entwined in his mother's hair.
Infinite

Spherical
loops.

Pain keeps arollin' all night long.

This could go on for ever.

Be patient, folks. This is a process for me.   "I can not forgive you." Tool.

"Naked and fearless" Tool again.

(DOWN HERE IF YOU"RE COMIN" FROM PAGE 44) LOL

who knows more about somethin' than I do. So there's no guarantee That I can come up with somethin'. But I'll give it a try.

47

XVIIII

loop

He bought it at a used-book store. <u>The Origins of Creativity</u>. People talking about how life experiences had led them to Amazing Discoveries. Haven't actually read it yet. Just thumbed through. Maybe he should read it now.

This could go on for ever.

Loops.
Spherical
infinite

Hands entwined in his mother's hair.

<u>The Sixth Sense</u> on TV.

XX

loop

Walking down Arrow Creek, naked, in the zone. Comes across a couple talking to a lone woman.
He sits on a rock beside her. Nice breasts. Large this time, but they still appear to be firm.
Curious, he begins to gently caress them. They are firm! What a surprise. She suggests that they
go off to the woods. Find a pebblely place. He feels something alongside his penis. A pebble?
She doesn't seem to notice. Maybe it's a seed.
Next day, while examining a salamander sunning on a rock, he hears his name called. Looks up,
there she is. Strange, can't remember saying a word to her. Must have though. She appears to be
with another man. He smiles and returns to the salamander.

Infinite
spherical
loops.

He may have a child he'll never know.

Hands entwined in his mother's hair.
This could go on for ever.

XXI

A woman's breast. Something between solid and liquid. Gelatinous! Very interesting. Curiosity
wins out.
loop

XXII

One more story about nudity, then he'll move on to something else. Maybe return to it later, there are more stories to tell.

He saw his first real-life, naked lady in a life drawing class in college. Late-bloomer. And to be honest, it was the reason he took the class in the first place. He had been a pretty good drawer before going off to school. Realistic trees. Then they tried to teach him how to draw. He tried it their way. Lost his touch. Now he couldn't draw a lick.

At one point, the woman wanted to tie her hair back. He had a hair tie in his pocket, remember he had let his hair grow long too. He nervously offered it to her. Still a juvenile in college.

Hands entwined in his mother's hair.
Infinite
Spherical
loops.

She took it.

This could go on for ever.

XXIII

loop

John Mellencamp (Originally. He went by Johnny Cougar), Kurt Vonnegut, David Letterman, he/me. All from Indiana. Abe Lincoln, he lived here! What is it about this place?

Hands entwined in his mother's hair.

Robert Wise! He thinks he's from Indiana. Maybe he'll do some research and see if he can come up with more names.

They hung the first white men for killin' some Injuns nearby in Pendleton.

Up until then they just let 'em get away with it!

What is it about this place?

Wilber Winkie (That's probably mis-spelled). A socialist who ran for President.

(Here's a honest-to-god story, folks: He wrote Winkie down because that's what was going through his head at the time. He knew it was wrong but he couldn't get Winkie out of his head. He comes home from work, sits in his chair, and starts thinking. He does let these random thoughts go through his head at times and all of sudden it came to him. Wilkie! Then he remembers what he wrote and he starts laughin' out loud. Honest-to-god. Laughin' out loud. Here's this kinda famous guy and I call him Winkie. I do have a wicked sense of humor! Honest-to-god. Laughin' out loud. It had been a long time since he had done that and it felt good. Back to the original story. Winkie! Whew.)

He really does need to research this thing!

{He remembers somethin' about how Indiana had the first public school system.}

(I really do need to research this thin'.)

(Let's try the version from the <u>Spawn</u> soundtrack now. That's the version he first heard.)

Andy Kaufman! I don't know if he's from Indiana, but he should have been.

(It's by both Filter and The Crystal Method.)(Trip like I do.)(Feel like do.)
This could go on for ever loops spherical infinite hands entwined in his mother's hair.

(

52

He just googled Andy Kaufman,
actually he googled Kaughman first,
And
he can't find
any
reference
to
Kaufman
ever
being
in
Indiana
on
Wikipedia.

Damn!

That
would

Have
been

a

nice
coincidence.
)
(But it did reinforce an idea he had goin'.
About havin' the author commit suicide before he finished this thingy.
Because this thought had crossed his mind.
I am getting a little carried away.
Nobody's goin' to get it!)

On with the show!

XXIIII

He thinks: You know, I think I might have had a poem published.

Back in high school, maybe grade school.

There was a contest and he wrote two.

One rhymed and one didn't.

He remembers part of the one that rhymed:

Indiana is my state.

It is a state that no other state can out-rate.

(He hopes it was grade school, not high school.}

He has no memory of the other one.

Just that he wrote two.

They picked one.

It was the one that didn't rhyme.

(Good, the alignment didn't get messed up.

or else the followin' wouldn't have worked.)

He keeps telling the teacher:

They should have picked this one.

It rhymes.

Infinite

(He thought poems were suppose to rhyme back then.)

(Can you see the rhythm here?)

{Listen to your heart beat.}

spherical

54

Loops.

A little The Crystal Method now: Do you trip like I do?

Sometimes he gets frustrated. It probably showed in that second suggestion he made. But he was wondering: I wonder if they got it yet?
Then he realized: Probably not that many people are familiar with how a sound wave looks on an oscilloscope.
So, in a more traditional manor, he'll take the time now to explain rather than make you wait.

It undulates. Goes up and down, up and down. Maybe he can see if they could print the preceding page horizontally so it might be more obvious. But then you'd have to turn the book sideways to read it.

So, look at the relationship between the beginning of the sentences. Some are long. Some are short. They go back and forth, back and forth. Or, if you just turn the book sideways, yeah, that would be easier, up and down, up and down.

To him, that's rhythm.

It's noticeable on a left justified page if you look at the end of the sentences. That's what turned him on to it. But in Western society, we're taught to go left to right. So, he turned it over in his head, played around a little, and thought:

Maybe I can right justify this sucker!

It might throw some people off.

Make them go:

What the hell?

But what the heck!

And if you had already gotten it and he ruined the thrill of discovery for you by making you read this rather long explanation:

He apologizes.

Wait to you see the next page!

One more thought first: He gets a kinda rhythm goin' in his head both when he reads and writes.
Do you do that too?

I want you to trip like I do.

Listen to your heart beat.

The music of your Universe.

{pound them knees}

(He's playing Babe Ruth's <u>First Base</u>.)

"And I laughed!"

Maybe I am starting to get too clever!

{Pound them knees!}

loop

This could go on for ever.

(<u>First Base</u> has a cover of one of Zappa's instrumentals.)

(<u>King Kong</u>)

XXV

loop

Here's this thought:

There's probably a black hole at the center of the known Universe.

And all these other black holes are just new Universes trying out new things.

Hands entwined in his mother's hair.

You'll probably think he is crazy.

Infinite
spherical
But it sounds good to him.

Loops.

This could go on for ever.

What do you think?

Be come a child again. Forget what you know and use your imagination.

XXVI

loop

Somehow he downloaded this during the madness:

Don't know how he did it.

He was just playin' around at the time.

"Like Humans Do (radio edit)" by David Byrne

He downloaded it a couple of times.

So, it plays a bunch of times.

He's listening to it now.

Hands entwined in his mother's hair.

infinite

He really can't remember how he did it.

Spherical

He really was just playin' around at the time.

This

loops.

Could go for ever.

XXVII

loop

Also durin' the madness:

He was sittin' at his computer, naked.

And he thought:

I really like being naked outdoors.

If someone sees me, they'll think I am crazy.

So, he just sits there at his computer.

Again,

I really like being naked outdoors.

If some one sees me, they'll think I am crazy.

So, he just sits there at his computer.

This loop goes on for ever

Finally:

What the hell!

I am crazy.

So, he runs outdoors naked.

In the snow.

Barefoot!

Hands entwined in his mother's hair.

Infinite

loops.

He thinks his neighbor did see him.

This could go on for ever!

It's in the way his neighbor looked at him while, deep in thought, he was outdoors, later, clothed this time, smokin' his pipe.

Spherical

But the neighbor didn't call the cops. Thanks neighbor.

The madness was really about his way of facing his fears.

XXVIII

loop

(John Mellencamp's <u>Uh Huh</u> in the background.)

loop

His mother's mom died when he was quite young. The story goes: His parents returned home
from her funeral and ended up taking him to the hospital. Meningitis. All he remembers from that
time is a toy barn, plastic cows, a tractor. At a reunion for his father's side, he sees a slide of his
father's mother sitting by his side at the hospital. What a surprise. He had been in intimidation of
her. Wasn't sure if she liked him or not. Before she died, he sent a print of a slide he had taken.
Water Lilies on a pond from their home town. Hoped she liked it.

His memories of his mother's father were of playing checkers. Old man with an afghan on his
lap. Those realistic trees he had drawn were beech trees in his grandfather's yard. His grandfather
was quite a character, trying this, trying that; his mother was born in a sod hut in Montana, before
buying a bunch of farm land in Northern Indiana. Nice black soil. Even predicted that man would
go to the moon long before we actually did.

To be honest. he doesn't remember that much of his father's dad. His father and grandfather had
owned a small grocery store. They use to give him candy.

He was not the only one intimidated by his grandmother. His grandfather and grandmother
moved to California, leaving his dad holdin' the bag on a store gone bust.

His grandmother never knew.

Hands entwined in her hair.

Hoped she liked that print.

Infinite

At the reunion after her death, he also heard that his father's mother was into photography.

Spherical

Maybe it runs in the family?

Loops.

This could go on for ever.

"If I could cry, I'd cry for everyone."

63

Gentle Giant said it first on <u>Octopus</u>. He's listening to that now, staring at what he just wrote. Don't cry. Time to go back to that dead-end job. And become the smiling imbecile they think you are.

(There's more to this story but I'm in a hurry.)

XXVIIII

Boy, he just got back from work and a bunch of things happened. A couple of paths to take. But he'll try to stick to the one that relates to the preceding sentence: And become the smiling imbecile they think you are.

(Tantric said it first: You seem to have control.)

When he makes statements like that, people usually say: You're just being too sensitive.

Or,

If they're in the psychiatric field:

You're being paranoid!

(Tantric again: You're being paranoid.)

Let him take some time to set this one up:

He works in a warehouse that handles music Cd's. He works in the department called Hits. It's the top selling Cd's. and kinda small compared to Catalog, where they keep all the Cd's still in print for the company they warehouse them for. So, they just put the boxes for orders on carts. The carts hold nine boxes. We have three sizes of boxes. One holds 30 Cd's, one holds 60 Cd's (you can actually get 64 in one if you play around a little. That's because it's sized to hold two 30 count boxes.) Then there's the big one, 120 Cd's (Again, because it's sized to hold 4 boxes, you can get more than 120 individual Cd's in it.).

So, the auditor (they don't audit all the orders, just some.) brings a cart to him 'cause (You know, there's this: I keep getting interrupted thing goin' on. That may happen later. Anyhow, he's typin' along and notices brings is capitalized. I don't remember doin' that. So, he goes back to try to make it a little letter. And he can't. The program won't let him. It keeps capitalizin' it. Finally, he makes it a little letter. The reason he don't like computers that much: The programers went in one direction and he wants to go in another. He had that problem with other students when he was studin' computer programmin', but maybe we'll leave that for later.)

Anyhow,

Where was I?

Oh, yeah.

65

He's
waitin' (It's doin it again with He's, it's doin' it here too. Listen. I know what I'm doin', you
don't have to do it for me.)

Anyhow,

(a little Alice in Chains now: Stomach hurts and I don't care.)

(It actually does some.)

{Actually, it feels more like my intestines.}

(But if you tell the doctors that, they think:)

(He can't tell the difference between his stomach and his intestines.)

{Well, yes I can.}

(I have a pretty good idea where my stomach is because I can feel it when I drink cold liquids.)

(And that's not where it hurts!)

(Down in a hole)

(Alice in Chains again)

(Maybe , He'll just listen to the music awhile.)

Anyhow,

to pack it. And the auditor says: There's two little boxes under the big boxes.

Maybe, he should point out that the boxes on the cart go:

Three across on one shelf,

Three across on a shelf below that,

66

And,

Three more boxes on the bottom.

They're all visible just by lookin' at it.

You know. He took the day off to get some things done that he's neglected (like bills) because
(The program is doin' weird things again!) (There, that fixed it! Just needed a space between)
and because.)
He'(Did it again! Damn! Forget about it and just keep typin') s been at the computer,
writin'.
And here he is sitting at the computer. Writin'. Listenin' to music. Thinkin'.
Maybe it's time to do those things. He'll be back later. He's got some of this written out on paper
so he should be able to pick up where he left off.

I actually am enjoyin' writin' this way!

Damn?

(What in God's name have you done.)

Alice in Chains.

{He keeps playing around with things.)

(Seein' how the words look on the page.)

(Really do need to do those other things.)

Bye now.

And I was just startin' to have fun.

Damn!

Back!

Goin' to have to scroll up to see where I left off.

Anyhow, (Still haven't gotten to those bills! He took Thursday off and this is Saturday.)

And when you pack

(Plan to finish this one, too.)

(Still writin' in here.)

This is as good as place as any to put the following: No I won't. I just scrolled down and it's already in here.

Take time out and go listen to Nirvana's "Dumb", please!

Maybe he should take the time now to tell you the connection he made between:

Are We Not Men!

These are the Laws!

And computer programming.

(Just watched an episode of <u>My Name is Earl</u>.) (It was about labels.)

(Seemed appropriate to this whole thing.)

(He was gettin' a little discouraged thinkin' about somethings.)

Anyhow,

(Plan to finish this one , too.)

Starting to tie up loose ends.

He's mentioned a couple of sexual escapades in a way that might make you think:

He thinks he's a stud.

Here's the way he looks at it:

Well, just before dozing off, he did hear Nancy mention that he was one of the few men to give her an orgasm. Seemed kind of strange at the time. He was pretty sure she had been with a lot of men.

But he's not sure if he's mentioned it yet; Ya, he did, it's toward the beginning, that he wasn't circumcised. And, he was kinda sensitive down there.

When he tried to have sex for the sake of having sex, there were a few times when he didn't even get it out of his pants before it happened.

Embarrassing!

So, he figured:

You know, when I kinda have this feeling of falling in love, gettin' into that zone thing, I'm better at it.

So, that's the approach he tried to take.

Hands entwined in his mother's hair.

Infinite

spherical

That's the approach he tried to take.

Loops.

It didn't always help then.

This could go on for ever.

71

There was a book by Roger Zelansky. Hope I spelled that right. might be legal problems
if I don't.

Anyhow, it was about the end of the world.

Had a kinda time warp thing goin' on.

He kept tellin' the same story over and over.

from different angles.

That's kinda what's happen' here.

don't know if I ever finished that book.

Hope that doesn't happen to you with this here thingy.

You might have noticed:
He's starting to use titles, instead of chapter numbers?

That's because he starting to write backwards.

Doesn't know what chapter it will be.

Originally, he thought:

(Dad just called. I keep getting interrupted!)

(Turns out he's late for lunch. Forgot to reset his clocks.)

(After lunch.)

(Most of the time he remembers his clocks are off a hour. But this time he was deep in thought, trying to compose this story in his head.)

{Stop trying to think ahead and just write!}

Back to the story.

Originally, he thought:

I'll put things in the order that I think of them. It will jump around because of that. But because this thing is as much about trying to illustrate how I think as much as anything else, I'll put them in the order I thunk 'em. So, he does that. Actually he had already made a couple of changes, but not many.

He writes this thing, looks at it and goes:

You know,

This kinda completes the circle.

Encloses

the infinite, spherical, loop thingy.

I wrote about this same experience in pretty much the same words.

So, maybe I should put it at the end.

So, he did.

And then he went:

I started this thing there, I always thought I might say more about it. Maybe it's time to do that now.

So, he started to write backwards.

Hands entwined in his mother's hair.

And guess what?

Infinite

That isn't the end now.

`Spherical

He's added even more to the ending. A coupla times.

Loops.

Are you getting a feel for that infinite, spherical, loops thingy yet?

This could go on for ever.

(There are still some things he wrote out on paper that he hasn't typed in yet. Maybe he'll go back to where he left off, and start from there again. Hope that things meet in the middle.)

This is the beginning of the end; my friend.
He thinks Jim Morrison said that first.
He's not certain, he's not playing any music at the time.
But he thinks that's how it goes.

How his relationship with Nancy ended:

He had gotten into photography. The reason he went to Arcosanti:

He worked with a fellow architecture student at a factory during the summer. Hi, Bill.

That student had been at Arcosanti.

They'd go get a couple of beers after work, talk.

School starts up again.

It was the early 70's, right after the Olympics in Germany.

They had used a poured, formed, not sure of the technical term, technique using concrete to build
some buildings there. Actually, they may have used fabric. It was a long time ago. He's not
certain now. Maybe they used both. He's not certain now.

They get an assignment:

Design a clubhouse for the local sailing club.

He let his imagination go wild. Wasn't even certain it was physically possibly to do what he had
in mind. But it looked like a sailboat to him. Even had a wave in front, a mound of Earth.

The client comes in:

Well, we have a clubhouse made outa block. If you want, you can knock out some blocks.

Despair!

I want to design a sailboat. I don't know if it's physically possible to do it. But I want to design a sailboat. And all we can do is knock out a couple of blocks.

Maybe this isn't what I'm suppose to do.

He remembers Arcosanti. Maybe I should go out there.

He talks to Bill.

Bill was trying to put together a slide show about Arcosanti.

Bill suggests:

Buy a camera. Take some photos and I'll use them in my slide show.

He knows absolutely nothing about photography, cameras.

So he asks another friend to help him buy a camera.

Pratikca. He thinks that how it's spelled. Maybe Praticka. No, that's not right. It's probably still around here somewhere. In pieces. It quit working and he tried to fix it.

Anyhow it was a nice camera.

(He's playing music now. The Commitments: Mustang Sally.)

Shutter release on the front, not on the top.

It felt comfortable in his hands.

They don't put them there anymore.

Anyhow.

(I don't know why I love you like I do.)

(I don't know why you treated me so bad.)

(Take me to the river.)

(Chain Of Fools.)

{There's that tangent thing going on again!}

{Sorry about that. Just tryin' to set the scene.}

Anyhow.

He takes some slides, gives them to Bill. Don't think he ever got them back.

He thinks:

I kinda like photography and I seem to be pretty good at it.

So, he takes a bunch of photography classes, from a bunch of different departments.

Architecture.

Art.

Industrial Arts.

Different approaches to the same thing.

(Try a little tenderness.)

A bunch of things he was taught to do he was already doing.

That whole divide it into thirds this way. Divide it into thirds that way. That's how you center a photograph. He was already doing that.

But he did learn some new things.

(Hey, Hey, Hey.)

Anyhow.

He thinks:

I might be doing this for a while.

Maybe I should learn how to fix a camera.

(At the time, he was academically trained, not mechanically inclined.)

So, he goes to school in Denver. National Camera Repair.

They did it mostly by mail, but he likes to ask a lot of questions, and they had classes out in Denver.

(They call me mister pitiful. That's how I got my fame.)

While out there, he receives a letter from Nancy. Don't know how she got his address.

(That's how I got my fame.)

She's in Georgia, teaching. Wants to see if they could keep their thing goin'.

He thinks:

They've got strip clubs here. I liked her body. Maybe, she could come out here and strip for a living.

(Very, very pitiful!)

So he writes back. Makes that suggestion.

(What would I give for a few more minutes.)

(Slip away.)

Anyhow.

A couple of years later, he hears she living in Cincinnati with her sister.

Gets her address.

Decides to make another road trip.

It's during his heavy drinkin' period.

(The Doors: The Soft Parade now)

(Tell all the people that you see, follow me.)

(You're life's complete.)

Not quite yet!

(Follow me down.)

(You got to follow down.)

(We'll be free.)

(I'm not afraid!)

(I'm goin' love you till the stars fall the sky.)

She's not home.

He starts drinkin'

Waits.

Drinks some more.

She's still not home!

So, he goes to a nearby bar. It's night by this time.

He's pretty drunk already.

He decides to call her.

She answers:

I don't have anything to say to you.

He replies:

I don't want to talk to you. I want to see you.

(Very, very pitiful!)

(It will be an easy ride.)

He goes back to her sisters house.

Passes out on a lawn chair.

A cop comes along.

Wonders what he's doin' there.

A black guy he's talked to a couple of times while waiting is there.

The black guy says:

He's been here all day.

The cop asks if he's got a place to stay.

The back of my truck.

(Much too easy rememberin' when.)

(I've got the running blues.)

(Don't fight, too much to lose.)

(Wishful, sinful, wicked youth.)

So, he finds his way back to the truck.

Passes out again.

(Can't escape the blues.)

The next morning, he awakes.

Looks out the back of the truck.

He doesn't have his glasses on.

So, all he sees is a dark shape.

Maybe it's Nancy!

He puts on his glasses.

Gets out of the back of his truck.

The shape is no longer there.

Hands entwined in his mother's hair.

(Petition the Lord with prayer.)

Infinite

(Listen to The Soft Parade)

Spherical

(Really! Go listen to it. He is.)

Loops.

This could go on for ever.

(Very, very, very pitiful, indeed.)

(Don't cry.)

(I'm glad that we came. I hope that you feel the same.)

(Carrying a heavy load.)

Delusions = distorted thinkin'

REM: <u>Everybody Hurts Sometimes</u>

Why he thought that dead-end job had killed his love of learning:

When he first started working there, he was kinda interested in learning everything they could teach him. Any job they would show him. And he did do a number of jobs at first. Bounced around a lot.

(Uriah Heep said it first: just out there, traveling in time.)

He doesn't like to take on very many responsibilities at once because those he does take on, he takes very seriously. He was driving a reach truck, putting away product. It was the busy season, and the more he tried to put away, the more there was. He started hurrying and he knew: the more I hurry, the more I fuck up. But he got in a hurry anyway. And he kept trying to tell them: take me off this damn thing. I'm going to fuck up!

He'll put this here: He wasn't brought up to cuss. It's just that he has been in a # of environments where cussin' was common.
(Not exactly in those words.)

They became a part of his Vocabularry. They slip out every now and then. And sometimes they seem like the best way to express yourself.
And he did fuck up! And they took him off the truck.

He got a job where he could do a variety of things; pick, pack, make boxes, keep busy with out having to do one thing and get in a hurry. And he thought: I kinda like this.

So, he'd go in for reviews and they'd say: Would you like to learn this job (Let's say: shipping)?

And he'd say: No.

(Remember, when he first started working there, he wanted to learn everything they would teach him.)

And the reason he'd say: No.

(Trapt: decisions to hide.)

(I can't give everything away. This is not where you belong.)

(Pound those thighs.)

Because they'd just stick him there for a couple of days, teach him a couple of things and send him back there to his old job. And they wouldn't put him back into shipping for a long time, probably because it was busy. And he learns by doing. It can take a while for him to learn things, especially if the way it is done don't make sense to him. And a lot of the ways they do things around here don't make sense to him. There should be a simpler way of doing this.

(Trapt again: Please help me 'cause I'm breaking down.)

(Take a deep breathe before attempting the following.)

So, he won't remember how to do what you want him to do because he didn't do it enough for him to remember how to do it because it didn't make sense to him to do it that way, there should be a simpler way of doing it and you want him to do it because we're busy.

(Breath deep 'cause here we go again!)

And he's goin' to havta ask a lot a questions because he won't remember how to do what you want him to do because he didn't do it enough for him to remember how to do it because it didn't make sense to him to do it that way, there should be a simpler way of doing it and ...

You're going to get mad at him for asking a lot of questions.

And he doesn't like for people to get mad at him, even though it often seems they do.

(It's all in the tone of voice.)

So, now he says: No.

Hands entwined in his mother's hair.

Infinite

SPHERICAL

loops.

I don't want to learn any more. The irony of it all!

This could go on for ever.

Delusions = distorted thinkin'

Has thought:

Maybe this thing will get me on the David Letterman show.

Maybe he'll want to talk to me before the show.

I'll go:

Dave, when I'm out there, ask me to tell you all I know about mass.

He'll look at me strange.

I'll say:

I don't want to spoil the punch line for you yet, Dave.

I'm on the show.

Dave asks me:

Tell us all you know about mass.

I'll go:

Have we got much time?

No.

Okay, I'll try to make this short.

I've gotta second Dave daydream but I gotta go to work now: Put in later, So it won't line up. Sorry about that.

Fuck it! I'll put it in now or else I'll go to work laughin' and they really will think i'm an idiot. We are on tape-delay, aren't we?

Dave, if you thought I planned this fucker to turn out like it did, Don't you think I woulda charged more than10$.

LOL

Actually, I got a third, but I'll keep it to myself just in case it actually happens.

LOL Got one about Oprah, too. lol

Let's see.

I know that weight is a measurement of an object's mass in a gravity field, specifically, the Earth's gravity field.

And I know that technically there is a difference between weight and mass because of that whole, it maybe weight-less but it's mass can still crush ya, thing.

84

And

That's it!
If you know that much about mass you know as much about it as I do.

The crowd stands up and applauds, laughing, whistling .

Ta Da.

He takes a bow.

Hands entwined in his mother's hair.

Infinite

Distorted thinkin' = delusions

Spherical

loops

This could go on for ever.

(Jim Morrison singing: Break on through to the other side!)

Someone, he don't know who, singing: I've seen better days, then the bottom dropped out.

A sign he made up and stuck on the wall above his computer at the beginning of the madness:

ONLY THE
MEDIOCRE
HAVE
NO
DOUBTS

(His script is so lousy even he can't read it. So he prints. And he usually uses all capital letters so he don't have to think about whether or not it should be capitalized.)

Hands entwined in her hair.

Infinite

(The time to wait is through, no time to wallow in the mire.)

spherical

loops.

This could go on for ever.

The Grateful Dead said it first: What a long strange trip it's been.

You may have noticed:

He's been pretty inconsistent. His writing style varies a lot. You're not suppose to do that.

Here's the way he looks at it:

I'm not a machine, I'm only human. Even when I try to do things the same way every time, I usually don't. Inconsistency is a part of being human. It comes naturally.

And he's worked in a couple of factories, worked with machines. And do you know what he found out? They're pretty inconsistent, too. They constantly need adjustment.

(Jim Morrison singing: I'd like to have another kiss.)

(Don't get freaked out! This is not a reference to his mother. He's not Anthony Burgess.)

Hands entwined in his mother's hair.

(Jim Morrison, again: Come on baby. Light my fire!)

(It's a preview of the next chapter.)

INFINITE

Spherical

loops.

(Some one said it before: Love having written, Hate writing. You know, this time, I'm kinda enjoying writing this way. Kinda afraid of what others will think when they read this. Who cares!)

This could go on for ever.

(Rare Earth's version of "Get Ready"now. It's pretty long. 21:29)

(They don't make 'em like that anymore.)

On the radio on the way home from work, Tom Petty singing: The waiting is the hardest part.

He went to see Ramona while she was in New Hampshire. She seemed happy. He didn't think a long-distance relationship would work, so he thought he ended it.

MY SPIRIT FLIES TO YOU

THE
BUDDHIST
MONKS

SAKYA TASHI LING

(In the background.)

(Pound those knees.)

He receives a letter from her, she's going back to that New-Age, clothing-optional, hot springs resort in Northern California with another woman.

He thinks: Is she asking me to go back there with her? I told her I had to stay here, work, pay off my truck, get back to the real world.

At the end, she says she had a one night stand, got pregnant, and had an abortion.

(Remember she was European, it wasn't her first.)

He thinks: Maybe this is what it's all about? She wants my sympathy and I told her I had to stay here, work, pay off my truck, get back to the real world. You're not showing any toward me!

If all she had written about was the abortion, her feelings about it, maybe she would have gotten his sympathy. But instead:

It pissed him off!

(Jim Morrison, again: take it easy, baby. Take it as it comes!)

She waited to the end to tell me what she was really writing about. She really doesn't know me at all.

So, he writes a very, very nasty letter, basically saying: Leave me alone! I told you I had to stay here. Work. Pay off my truck. Get back to the Real World.

88

(Ten years earlier he would have gone with her: but that was then and this is now.)

During the madness, he thinks: it only costs 42 cents to send a letter, she was looking for a New-Age community to live in, last he knew, she was heading to that New-Age, clothing-optional, hot springs resort in Northern California.

So he sends this out there:

Ramona

I'm taking a chance that you are still at HARBIN HOT SPRINGS. Taking A CHANCE THAT You STILL REMEMBER ME. BUT HERE IT GOES.
I APOLOGIZE. PART OF ME WANTED TO GO WITH YOU. PART OF ME NEEDED TO STAY. I WENT WITH MY NEED INSTEAD OF MY WANT.

IF YOU FORGIVE ME, I CAN BE CONTACTED AT: MIKE BARDEN

— — — ---

‒‒‒‒‒‒‒ ‒‒ ‒‒‒‒‒‒‒‒‒

{ He thought it might not be a good idea to have his actual address published.}

(Jim Morrison said it first: too late.)

infinite

It came back: RETURN TO SENDER
Address Unknown

spherical

That's why he can be so accurate as to what he said. He simply copied it.

Loops.

And he did use both little and capital letters.

This could go on for ever.

OK, this is what I was referrin' to on page 69. The reason I wanted to make sure this was in here:
 If this fucker does fly, you might think: What an imagination he's got!
 Well folks, I'm lookin' at the envelope with the postal stamp on it right now.
 And if this fucker does fly, I'm goin' to keep it in my back pocket to show ya.
 "Trip like I do"
 "Trip **like I** do*"* Many Many Paths To Take I really am channelin'   Hello
                                                                                      now
                                                              Frank!
                                                                        And
                                                        Tim!
                                                      And
                                                Jim!
                                          Either
                                      One!
                                At the speed of light><
                                                  A
                                              photon
                                          Awe!
                                      My
                                    Spirit
                                    Flies!
                                          <>

Honest-to-God folks: Laughin' out   Rock out   And roll with it

Jim Morrison singing: This is the end; my only friend, the end.
Well, we're not quite there yet, but we're getting closer.

He thought:

You know, maybe I am doing that Jonathan Winters, Robin Williams thing again. But way back then, the people he was with let him go with it. They found it amusing.

Poofing the donuts?

He does it now and he keeps getting interrupted. Others try to impose their thoughts on his.

Jeff, actually, I've already contacted a small publisher. I was goin' kinda crazy at the time and I didn't know why it was a good idea, I just did.

He found himself writing again and, before, he had just thrown it away! And he thought: Maybe I shouldn't do that this time!

He writes about what kinda makes sense at the time and his thinkin' keeps changin'. That paper he entitled: Random Order And Infinite Variability. At the time it kinda made sense but he's not so sure about that any more. The reason: Black Holes!

The reason they're called black holes is that they're so dense, their gravity fields so intense, that light can't escape from them. And it was thought that nothing can go faster than light. The speed of light is sorta a boundary condition of the known Universe. If you go faster than the speed of light, ya kinda have to go into another Universe and how do you know the Universe you come back to is the Universe you left. A whole new set of questions.

(Tool: Wise men may not be sober.)

{Trust me.}

There was a TV show built on this premise a long time...

(You're going off on a tangent again.)

(Sorry about that. Just trying to connect the dots. Set the scene. There is a, well, for you to understand this, maybe, I should try to explain that, kinda thing going on.)

So, if the speed of light is a boundary condition of the known Universe, and light can't escape the gravity field

(Things keep poppin' up on the computer. No, I don't want to update that now!)

of a black hole, then nothing should be able to escape the gravity field of a black hole. And guess what? They've found certain types of radiation emanating from black holes and that doesn't make any sense. So, the idea that what we see, what we pick out, are the things we can make sense out of may explain some things but it doesn't explain everything.

Hands entwined in his mother's hair.

Not everything, at least.

Infinite

spherical

Loops.

This could go on for ever.

(Tool again: Let it go!)

(Pound those knees!)

Inch by inch, step by step

(Nonpoint this time: Trip inside your mind.)

(I'm goin' take you on a mind trip.)

Back to that definition thing again!

When the psychiatrist told him he had schizophrenic affect which was actually months after he had been taking Geodon to combat it and if you allow him to go off on this tangent a bit more: The psychiatrist keep saying we haven't reached the therapeutic dosage yet and kept upping it. The last time the psychiatrist prescribed it for him, it took two prescriptions.

First thought:

It's fifty dollars a prescription! I can't afford to spend $100 on one drug! I'm taking a bunch.

So, he never filled it.

Then he thought some more:

It doesn't make sense for a therapeutic dosage to require two prescriptions!

Anyhow, after the psychiatrist told him he had schizophrenic affect, like he said before, he knew the drug part didn't fit, so he decided to google it. He knew schizophrenic meant delusional but he wasn't certain about the affect part. And do you know what he found for an explanation?

Flat emotional response.

And Dr. D. If you had given him that explanation, he would have agreed with you.

He did have a rather flat emotional response at that time.

But he also would have said:

It's not because of any drugs he had done.

Not because he had these seemingly peculiar thoughts.

Not because he was a loner.

It was intentional.

When he was emotional, he tended to be very emotional.

And, either, because of the way he expressed his emotions, or, of the emotions he expressed, remember that whole woman he stalked? thing, people tended to find his emotional responses inappropriate.

So, he decided to put a lid on them.

Hands entwined in his mother's hair.

During the madness, he started opening up again, be a little more lovin'.

But again, now he's back to the Real World.

So, who knows what will happen?

(Us and Them)

{Ordinary men, men, men}

(Pink Floyd, if you didn't recognize it)

{Up and Down}

(In the end it's only round and round.)

Infinite

There were a couple things he's wondered about in that ordeal with the psychiatric system. He always wondered: Are they trying to help me accept myself or trying to make me fit in? Do they have to be the same? Just because these are my experiences and they're outside the realm of yours, does that make them delusions?

The therapist said: You really are a non-conformist!

He replied: I'm not trying to be. I'm just doing what makes sense to me.

(the dark side of the moon.)

(All that you see, all that you taste)

(All you create)

94

spherical

(Try a little Chris Cornell, now: <u>Carry On</u>)

Discontinuation Syndrom?

It's real. Google it. Just a fancier way of saying withdrawals. Which is what he looked up. He knew something was different that time he had the impulse to take all the drugs he had to get rid of the pain.

And maybe, part of the time, the problem was that he didn't put as much emphasis on time as others do. Put quite as much distinction on past and present.

He had been asked: Have you ever thought about suicide?

He'd reply: Yes.

But, actually it had been a long, long time ago.

There was a time when, if he didn't have a girlfriend, a woman, a significant other; then yes, he did feel like no one loved him.

But that was then and this is now.

He knows his Dad loves him, his sisters, his nieces and nephews, probably the whole extended family. And even though she's dead now, if there is an afterlife, his mother still loves him.

loops

This could go on for ever.

{Don't cry}

They still irritate him at times and he irritates them at times.

(It feels like I don't have to worry at all)

(Chris Cornell)

{Find me for ever.}

(Another empty page. Damn!)

PrePare Your Self You'Re about To Go into AnoTher
blacK Hole

Almost to the end.

loop

The thought just struck him: You know, I do have this kinda wicked sense of humor. I do love irony, the idea that these two things should be total opposites and somehow a connection has been made between them. That scene in <u>A Clockward Orange</u>. Yeah, it is pretty violent. You should be disturbed by it. But somehow, by matching it with this innocuous song, to him at least, its not so disturbing. The song itself, <u>Singing In The Rain</u>. That's kinda ironic! Most grown-ups hide from the rain, they don't like getting wet. And here's Gene Kelly, dancing in the rain, stompin' in the puddles. Just like a child again.

A lot of times, when he says something: yeah, I am kinda serious. But I'm also trying to have some fun here. I kinda see a joke here. This really shouldn't have happened like it did. What a surprise! Something was wrong. How strange! Why did it bother me so? Awe. Now days. It seems like people only see the serious, and they grow concerned. And because of that, he was starting to take himself too seriously. He wasn't having fun anymore. Become a child again. Have some fun. Experiment. Experience.

The reason he liked the road so much. He could go places where people didn't think they knew him, and because of that, he could be who ever he wanted to be. He didn't have to try to fulfil their expectations.

Maybe, he was starting to regain that sense of the ironic after all.

Hands entwined in his mother's hair.

infinite

Spherical

loops.

The irony of it all!

This could go on for ever.

97

One last thought before the end.

loop

Before this whole thingy ends, he'll try to explain that Observer alters the Observed which alters the Observation which... thing at the beginning. Remember, he said it could take a while, so please be patient.

It's about the act of observing.

We'll start out with the concept of sight. He doesn't mean to be politically incorrect, again, like he was by mentioning he was the only native English speaker in that class. He does realize that some people can't see. And he really, really, didn't mean to offend anyone that time, either. It was a kinda botched demonstration of a voice recognition program that got him thinking about ambiguity in the first place. Back then, voice recognition was rather primitive. The program was taught to recognize only one speaker and that speaker had a cold.

(You're going off in a tangent, again.)

Sorry about that. Sometimes it seems like there's away to connect just about everything.

So, if you can't see and you're reading this, please forgive him.

What happens when you see something goes kinda like this:

You take a photon of light, bounce it off an object, that photon goes and hits the back of your eyeball, excites some nerves, which fire off an electrical impulse, which goes to your brain and your mind says: I see an object. But what you're really seeing is a photon of light that was reflected off an object. What you actually see is a reflection of an object, not the object itself. It's just easier for your mind to think: I see an object, rather than: I see a reflection of an object.

So far so good?

Next step involves mass. You kinda can think of weight. Weight is just a measurement of an object's mass in a gravity field. Specifically, the Earth's. The Earth pulls on the mass of an object, you stick a scale under that object and, voila, you've got weight. Technically there is a difference, but it really only comes into play when you're out into space, out of the Earth's gravity field. That whole, maybe it's weight-less but it's mass can still crush you, thing. So, for the purpose of this explanation we'll just use weight. The weight of a photon of light is very, very small. Fact is, he kinda remembers that there might have been some question about whether or not a photon of light has any weight at all. But he's not really up on the current thinking, so we'll just say that the weight of a photon of light is very, very small. In the classical world of physics, the real-world that we all live in, the weight of the object being observed is very, very large.

98

Let's try some pool balls.

You strike the cue ball, which hits the object ball, and both of them kinda take off. That's the quantum world. If you use a marble instead of a cue ball, the object ball might move a little but not much. That's the real-world.

So, bouncing a photon of light off an object in the real-world has an effect on the object but since the difference in weight is so great, the effect is considered unmeasurable, negligible, ignorable.

Still with me?

Now we get to the quantum world. Back to the pool balls. The cue ball has the same weight as the object ball, so the object ball takes off. You can see it move. In the quantum world the weight of the object you use to bounce off an object; it doesn't have to be a photon of light, an electron will do, is about the same weight as the weight of the object being observed. The effect of bouncing one object off another in order to see that second object is now measurable, no longer negligible, can't be ignored. The act of observing an object changes, alters, that object. By observing an object, the observer alters that object.

So, he kinda played around with that idea a couple of times, turned it over in his head, and came up with the Observer alters the Observed which alters the Observation which... thing. Like he said before, there's probably a better way of putting it. He used all nouns. There probably should be a verb in there somewhere. It is a process, after all. But for now, that is going to have to do. He wants to explore new things, not re-hash old.

Hands entwined in his mother's hair.

Infinite

Let's see, The Observing alters...

Spherical

Loops.

This could go on for ever.

99

The End

loop

He wasn't exactly what you would call a linear thinker. He didn't try to figure out all the details at once. He'd think about one thing for a while. Decide: that's really all the farther I need to go with that thought for now. Maybe I'll think about something else for a while. So, he'd think about something else for awhile. Take that as far as it needed to go. Decide: Maybe I'll go back to what I was thinking about before. Maybe I'll go on to something else. Things kinda came to him in spurts.

He had always known that his relationship with Ramona had been kinda strange. He had only known her for six months and was actually with her for about, let's see, three months. But he had felt closer to her than any woman before or since. Had come up with a couple of possible reasons for that but none of them really seemed to explain it all. It wasn't until he was writing this down that he really understood. In those three months, she had shared more with him about her life, her thoughts, her feelings, than any other woman before or since.

Has thought: Maybe I should look at this thing as a first draft. Go back and organize things in a more traditional manor. Make things a bit more accessible. But he wasn't trying to write a best-seller. Didn't care if things were a little hard to follow. The books he had liked the most were the ones that he didn't always understand what the author was talking about at first. But he was kinda intrigued. So he'd think: Maybe I'll read a little more, see if things start making a little more sense. And to his amazement, they did. Maybe that's what the author was talking about back there. Amazing Discoveries. Awe. How strange. What a surprise! Become a child again!

The author hadn't forced him to see things his way, but had allowed him to discover what was going on.

Infinite

Maybe he'll go back, organize things in a more traditional manor.

Spherical

But at this moment, he doesn't want to.

Loops.

Not really.

Hands entwined in his mother's hair.

100

Find his voice, that inner child.

This could go on for ever.

Appendium

loop

He had thought about going back to that New-Age, clothing-optional, hot springs resort in Northern California. Told this to his therapist.

The therapist asks:

Did you feel at home there?

Not really.

Home wasn't a place. It was in his heart.

(don't cry)

Has thought about Arcosanti. Maybe he should go back there!

Find a mound of Earth.

Stand.

Naked.

In a thunderstorm.

DARE the lightning to strike!

Hands entwined in his mother's hair.

Is the pain gone?

infinite

Almost.

Spherical

If others see him, they'll probably think it is weird.

loops.

Singing in the rain.

this could go on for ever

{a really deep breathe before attempting the following}

the observing alters the act of observing which alters that which the act of observing is observing which... no, i think I'll stick with the observer alters the observed which alters the observation which... the universe is a series of infinite spherical loops this really really really really could go on for ever !? hands entwined in his mother's hair

ta da

I knew you could do it if you really tried.

Take a bow.

That's all, folks.
For now at least.

Let's try Pink Floyd's <u>The Dark Side Of The Moon</u> again.

I think I've been mad.

I've always been mad.

Even when I'm not.

TA DA

{I really need to finish this damn thing!)

(This really, really really, really could go on for ever!)

We're goin' to enter a black hole now.

infinite
spherical   loops

103

```
        expanding              .              contracting

                like                    a
                   heart        beat
                        music
```

His world.

Welcome to it.

I hope you had fun.

He did.

Don't cry

(U2 this time: I still haven't found what I'm looking for.)

{This really, really really really, really could go on forever?}

And you give yourself away!

He's said he doesn't like to be the center of attention.

And

If this thing takes off.

He might draw attention to himself.

The IRONY of it ALL.

(I really do need to finish this thingy.}

<u>Random Order And Infinite Variability</u>

(That's starting to make sense again.)

(At least the title is.)

{Maybe I'll make one last attempt to find it.}

Things just keep on clickin'.

WELCOME TO MY WORLD PRAYER DISTURBED PART> HIS MIND IS
RACIN" TRYIN' TO CONNECT THE DOTS&LIGHT BULBS ARE POPIN'
OFF ALL OVER THE PLACE>THERE ARE MANY MANY PATHS TO TAKE

He'll try to start as many as he can: thank you VIRginny and Jim: you're my inspiration for this"A little EVIL inside': Disturbed again   !Pound Them Knees!   I'm not"

"Even when

A person^ he shared a house with in San Fran. How Ya doin' Jim? Did ya make it?
I didn't like all of Jim's stuff but he had this one collage that I'd just stare at

Pink Floyd
tryin' to connect the dots.

the
reason
he didn't
keep up
with
old
friends:
they
were
still
followin'
their
dreams.
he
had abandoned
his
wasn't
sure
what
they
had
in
common
anymore

Collective Soul
"Will love be there"
"Heaven let your light shine
down"

Most of anything is empty space.

Hello, Diane, you're in Earl's story.

Common folks vs. experts
Accordin' to the laws of probability:
The same 6 numbers picked in the last drawin'
have the same chance of bein' picked in the next drawin'
As any other 6 numbers
Most common folk would avoid those same 6 numbers

My subconscience
led me to this
writin'. There may
of clarity.

have been a moment
I'm goin' let
it go and let K_arma
Go the rest of u

How ya doin' Kurt?

Thanks for the
encouragement
Kurt

lo.op

A Dr. Bolin use to tell him about Robert Frost
He wondered why? He's not a poet. and he had given up
on bein' famous

Heard that einstein died unhappy.
Tryin' to prove his theory of relativity with math.
Heard that einstein was a lousy mathematician.
Theorist. he was great, but mathematician. no.
He shoulda left his theory alone. And let some one else Figure it out.
even if i'm right about powin' a hole in the laws of probability which i'm not sure i am.   that's as far as i can take that thought
Someone else is goin' to have to pick up the ball and run with it.

He really, really has gone too far this time LoL
Don't cry he already did it for ya. at work.
there may be one more thing he can learn from einstein.

Most of the time when
people tell him what to do
he wonders why
he already knows that
Can't figure out why? you're tellin
him that.

Gotta story
about Bad
I'd like
to share
with
you

I'm channelin' Frank and God knows who else?
This is a scam. He just changed the #
of pages he's submittin' to createspace.
Now he gets a whole $2.34
per copy. He really is goin'
to get rich off this f-----. LOL

Earl as in: at work not as in: "My Name is Earl" I told you, you would be in my book. But your story isn't. sorry about that.

Maybe now he knows why? Dr. bolin was tellin' him that.
Head in the clouds again>at the speed of liight<everything
is
relative

Jim was this multi-media, starvin' artist. And very very moody. Some times an asshole. sometimes very nice. he usually had to get stoned to be this way.
the second Jim   Wrote a letter to this asshole. Don't knoe how he got his address. Didn't think they were that close.
hands infinite entwined spherical in loops his . mothers this hair could go on for e

ver   and ever and ever anyhowfor lo.op so I'm goin let this fucker fly now

Waitin' to hear from ya. Bye now.

106

(I really have thought: this is how I want to end this thing.)

(Then sometin' happens to inspire me to add more.)

(Well, I just got back from havin' lunch with Dad.)

(He's told me in the past that he has been intimidated by my intelligence.)

(I don't like people to be intimidated by my intelligence.)

(Because, quite frankly, I'm not that sure I am.)

(There is always someone who knows more about somethin' than I do.)

(Like I said, during the madness, I started to open up a bit more.)

(And I've been talkin' my head off.)

(About the book, what ever comes to mind.)

(Most people look at me with concern.)

(But here's what I learned today about my Dad.)

(3/24/2009)

(He gets me.)

(I really like this guy.)

(All this time I've been a real asshole.)

(Don't cry)

(Anyhow = loop)

(Time to pay them bills.)

"My Spirit Flies To You"

at
the speed
of
light

www.ingramcontent.com/pod-product-compliance
Lightning Source LLC
Chambersburg PA
CBHW081136170526
45165CB00008B/2690